Lecture Notes in Earth Sciences

39

Editors:
S. Bhattacharji, Brooklyn
G. M. Friedman, Brooklyn and Troy
H. J. Neugebauer, Bonn
A. Seilacher, Tuebingen

Sven-Erik Hjelt

Pragmatic Inversion of Geophysical Data

Springer-Verlag
Berlin Heidelberg GmbH

Author

Sven-Erik Hjelt
Department of Geophysics, University of Oulu
P. O. Box 400, SF-90571 Oulu, Finland

"For all Lecture Notes in Earth Sciences published till now please see final page of the book"

ISBN 978-3-540-55622-0   ISBN 978-3-540-47263-6 (eBook)
DOI 10.1007/978-3-540-47263-6

Typesetting: Camera ready by author

32/3140-543210 - Printed on acid-free paper

PREFACE

The systematic development of geophysical quantitative interpretation opened up for a wide audience with the appearance of the famous book by Grant and West in 1965. The revolutionary papers of Backus and Gilbert soon followed. Computerized inversion techniques were about to start to develop. In 1973 I had spent more than a year working on computer implementation of magnetic inversion using non-linear optimization. This topic had become popular in the geophysical literature. The optimism over making the interpretation procedure automatic and "objective" soon turned into more realistic concepts of interactivity, the increasingly important role of the human interpreter - the geophysicists. The computer was really only a tool.

Many, even now, fundamental papers on geophysical and more general inversion appeared at that time, many papers repeating the same basic principles over and over again. It was a time for unifying ideas and developing an introductory course on the general principles of geophysical inversion. My first lecture series "Geophysical Interpretation Theory" was given at the Helsinki University of Technology in the fall of 1973. It concentrated on non-linear inversion with optimization and the technical aspects of writing geophysical computer codes.

Generalized inversion and linearized problems started to become fashionable at the same time. After having moved to Oulu in 1975, introducing the same course there, I had a chance to visit Aarhus in 1976. A group of young enthusiastic geophysicists, Laust B. Pedersen, Kurt Sörensen, Hans-Kurt Johannesen and others had been exposed to generalized inverse theory by Professors Ulrich Schmucker and Peter Weidelt. Via the discussions in Aarhus, via lecture notes which Kurt and later also Peter kindly made available, generalized inverse theory become a standard part of my lectures, too, both in Oulu and Helsinki. They form the solid background for the text in Chapter 3.

Many excellent books have been written (also primarily for a geophysical audience) since 1965. Many books appear to get so involved with deep mathematical aspects of inversion theory that the main aim of geophysical modelling and interpretation, to explain the wide and complex phenomena and structures of Nature itself, seems to fall aside. Therefore, I believe there is still room for an introductory text, which puts the emphasis not so much on mathematical elegance and completeness, but on the very basic concepts of inversion, not forgetting the human aspects, and, what is more important, on describing how and why geophysicists have selected and used various techniques, as well as the limitations and pitfalls. In a single phrase, a text on **pragmatic inversion**.

I am grateful to a great number of colleagues who, over the years, shared my interest in geophysical inversion, many of them coming from the electromagnetic induction community. The feedback from students has kept the course on inversion alive and under constant development. One young fellow, Jarkko Jokinen deserves special thanks for his enthusiasm in helping to scan and manipulate the figures of my lecture slides, thus presenting a form suitable for electronic redrawing.

Oulu, Finland, 15 January, 1992. **Sven-Erik Hjelt**

CONTENTS

Chapter 1	INTRODUCTION	
1.1	Definition of Inversion	3
1.2	Geophysical Models	10
1.3	Model Parameters	16
1.4	Fitting Model Fields to Data	17
1.5	Factors Affecting Inversion	18
1.6	Computational Aspects	25
1.7	Psychological Aspects of Inversion	29
1.8	What then is Pragmatic Inversion?	31
	References	32

Chapter 2	INTERPRETATION USING NOMOGRAMS	
2.1	Characteristic Points	35
2.2	Nomograms and Their Use	38
	2.2.1 Gravity Nomograms	38
	2.2.2 Nomogram for Interpretation of VLF Resistivity Data	41
	2.2.3 Nomograms for Magnetic Dipole (Slingram) Profiling	42
2.3	Computerized "Nomograms"	44
2.4	On the Accuracy of Characteristic Point Inversion	46
	References	46

Chapter 3	LINEAR PARAMETERS	
3.1	Defining the Linear Problem	51
	3.1.1 Genuine Linear Problems	51
	3.1.2 Discretization of the Linear Model	52
	3.1.3 Linearization	53
	3.1.4 Iterative Approaches	55
3.2	Fitting Linear Models to Data	57
	3.2.1 Systems of Linear Equations	57
	3.2.2 Exact Fit	60
	3.2.3 Least Squares (LSQ)	60
3.3	Iterative Solutions: The Method of Gauss-Seidel	61
3.4	Generalized Inversion	63
	3.4.1 Singular Value Decomposition (SVD)	64
	3.4.1.1 Zero Eigenvalues	66
	3.4.1.2 Zero Eigenvalues, a Least-Squares Approach	67
	3.4.1.3 Why are Small Eigenvalues Dangerous?	69
	3.4.2 SVD of a Rectangular Matrix	72
	3.4.3 Geophysical Examples of the Use of SVD	73
	3.4.3.1 Gravity	73
	3.4.3.2 Earth Density Model	78
	3.4.3.3 Magnetometry	81
	3.4.3.4 Magnetometry, Theoretical Plate Models	82
	3.4.3.5 Electromagnetism	90
	3.4.3.6 Seismology	94

		3.4.3.7	Oceanography	96
		3.4.3.8	Joint Inversion	97
	3.4.4	Ridge Regression		100
3.5	The Backus-Gilbert Approach			101
	3.5.1	Definitions		101
	3.5.2	Parameter Trade-Off		103
3.6	Geophysical Tomography			104
	3.6.1	Seismic Tomography		105
	3.6.2	Tomography and the Radon Transform		107
	3.6.3	Example on Seismic Regional Tomography		110
	References			112

Chapter 4   NON-LINEAR PARAMETERS

4.1	Definitions			117
	4.1.1	Objective Functions and Norms		118
		4.1.1.1	Exact Fit	118
		4.1.1.2	Least Squares	118
		4.1.1.3	Robust Methods. The Minimax Solution	119
		4.1.1.4	Weighted Objective Functions	119
	4.1.2	Properties of Objective Functions		121
	4.1.3	Constrained Optimization		127
		4.1.3.1	Penalty Functions	127
		4.1.3.2	Lagrangian Multipliers	128
	4.1.4	Stopping Criteria in Iterative Optimization		129
4.2	One-Dimensional Optimization			131
	4.2.1	Golden Cut		132
	4.2.2	Fibonacci Search		133
	4.2.3	Speedup by Parabolic Fit		134
	4.2.4	The Secant Method		136
	4.2.5	The Gradient Method		136
		4.2.5.1	On the Instability of the Newton Iteration	137
4.3	Multidimensional Search			138
	4.3.1	Direct Search		142
		4.3.1.1	Sequential Search (Search by Parameters)	142
		4.3.1.2	Hyperparabolic Fit	142
		4.3.1.3	Pattern Search	144
	4.3.2	Multivariate Search		147
		4.3.2.1	The Simplex Method	147
		4.3.2.2	Steepest Descent Methods: the Gradient Method	149
		4.3.2.3	Steepest Descent Methods: the Method of Conjugate Directions	150
		4.3.2.4	The Levenberg-Morrison-Marquardt Algorithm	152
	4.3.3	Random Search Methods		153
4.4	Examples			155
	4.4.1	Magnetic 2D Profiling		156
	4.4.2	Gravity Inversion		165
	4.4.3	Seismic Refraction		166
	4.4.4	Magnetotellurics		167
	References			171

Chapter 5   MAXIMUM LIKELIHOOD AND MAXIMUM ENTROPY
    5.1   Introduction and Definitions   177
        5.1.1   Probability Density   177
        5.1.2   Measure of Information   179
    5.2   Probabilistic Formulations of the Inversion Problem   181
        5.2.1   The Principle of Maximum Likelihood   181
        5.2.2   The Maximum Entropy Method   184
        5.2.3   Bayesian Estimation   186
        5.2.4   Some Useful Expressions for the Probabilistic Approach
               of Inversion   187
            5.2.4.1   Normal Distribution   187
            5.2.4.2   Smoothness   188
            5.2.4.3   Weighted Least Squares   189
            5.2.4.4   Parameter Boundaries   189
            5.2.4.5   More on a Priori Information   190
        References   192

Chapter 6   ANALYTIC INVERSION
    6.1   General Principles   197
        6.1.1   Combination of Components   197
    6.2   Examples of Analytic Inversion   200
        6.2.1   Magnetometry   200
        6.2.2   EM Magnetic Dipole and an Infinitely Well Conducting
             Half-Plane   207
        References   211

Chapter 7   ADVANCED INVERSION METHODS
    7.1   Functional Analytical Methods   215
        7.1.1   Methods of Singular Points   216
        7.1.2   Method of Tightening Contours   216
        7.1.3   Method fo Finite Functions   219
    7.2   Continuation of Fields   222
    7.3   Migration   226
        References   226

Chapter 8   ERROR ANALYSIS
    8.1   Introduction   231
    8.2   On Linearized Errors   232
    8.3   Minimal Error in Least-Squares Inversion   233
    8.4   Correlation Between Parameters   234
    8.5   Advanced Concepts of Error Analysis   240
        References   243

Chapter 9   PARALLEL COMPUTATION IN MODELLING AND INVERSION
    9.1   Parallelism in Geophysical Problems   248
    9.2   Forward Problems   249
    9.3   Inversion by Optimization   250
    9.4   Large Systems of Equations   253
        References   254

SUBJECT INDEX   257

Chapter 1

INTRODUCTION

"When you have eliminated

the impossible,

whatever remains,

however improbable,

must be the truth."

(A.C. Doyle: A study in Scarlet)

CHAPTER 1

INTRODUCTION

1.1 DEFINITION OF INVERSION

Geophysical measurements are not done for the sake of art only. The ultimate goal is to solve some well-defined geological, tectonic or structural problem. For this purpose the data, the measurements have to be interpreted, that is translated into a physical model of the subsurface. Geophysical measurements depend on and are actually designed for variations of physical properties of the bedrock, its structures, minerals etc. Geophysical interpretation - or as it is more often called today - inversion of geophysical measurements can be therefore defined as the construction of a physical model of the subsurface. The physical model, depending on whether it is based on densities, magnetic properties, electrical conductivity or differences in seismic velocities often differs from the distribution of geological minerals. Models based on different physical properties may thus differ quite considerably, since they describe the subsurface variation of different physical properties. However, the models, more often than not, complete each other and a joint interpretation or complex interpretation, as it has been called especially in eastern European literature, will lead to the best description of the Earth.

This book tries to describe some of the most important common features of inverting different geophysical data sets. At all stages the emphasis will be on the practical, pragmatic, aspects of the interpretation process. For many practical purposes the virtue of a good inversion system or algorithm is its interactivity and speed. In prospecting interpretation it is expected to be as operational as possible. Being at its best an operative, an interactive data inversion system enables geophysicists to direct the next steps in a prospecting campaign. If a geophysicist equipped with a proper inversion system can bring waiting costs for drilling rigs down, the resources put into the development of such a system will pay its costs back quickly -- and with high interest.

Fundamentally there is no difference between approaching various sets of geophysical measurements. Consider a rectangular coordinate system, where traditionally z is positive down into the Earth, and x along the profile. One can speak of following subdivision:

Type of measurement	Inversion produces	Dimension of model
a) sounding	$param(z)$	1D
b) profiling	$param(x)$	partially 2D
c) arrays	$param(x$ and $y)$	2D horizontal
d) combination of a and b	$param(x$ and $z)$	2D cross-section
e) combination of a and c	$param(x, y$ and $z)$	3D

Some definitions of 2.5-dimensional models have been introduced in the literature, but their use is not recommended in order to avoid confusion with the newly risen concept of fractals. Another source of confusion are the different definitions of 2.5-dimensionality for different geophysical methods. In potential field theory, a 2.5-dimensional model means a two-dimensional model with constant cross-section, but with vertical ends at finite distances along strike. More often this definition is restricted to the situation only when one deals with the profile located symmetrically with respect to the ends of the body. In electromagnetism a system with a three-dimensional source field above a two-dimensional Earth structure has been called 2.5-dimensional.

A further dimension is added to geophysical inversion, when the joint inversion of measurements taken with different geophysical methods are combined. Genuinely time-dependent phenomena, such as hydrodynamics of water, air or the ionosphere, convection circulation in the mantle, are not considered in this volume.

Geophysical inversion methods can be divided in many ways. One possibility - followed in this book - is to use the scheme given in Table 1.1. There is considerable overlap between the methods and no generally accepted division exists. In addition a wealth of filtering methods have been designed for transforming the measured data into more suitable form for interpretation. Such

Table 1.1 Division and properties of geophysical inversion methods

Method group	Method	Subdivision	Comment
Qualitative methods			Overview of a research area
Quantitative methods			
	Special points	Nomograms for quick interpretation	
	"Rules of thumb"		
	Curve fitting	Visual comparison	Model curves
		Mathematical	Parameters either linear or non-linear
		Automatic Interactive	
Special methods			
	Analytical		
	Function theory		
		Continuation of fields Migration Finite functions	
	Statistical	Average source distributions e.g. total mass in gravimetry MEM = maximum entropy ML = maximum likelihood	

techniques vary from method to method and are of greatest importance in qualitative interpretation. Typically these transformations are filtering operations with aiming to emphasize or de-emphasize certain variations in the data. Upward and downward continuation of potential fields (magnetic and gravity) data are good examples, where in the first case long wavelength anomalies are emphasized and in the latter case the short-period variations dominate.

In magnetometry field data sometimes are transformed to the pole in order to produce a more symmetric anomaly map which may be interpreted more easily. The background idea is, that a magnetized body at the magnetic pole will always produce a symmetric anomaly pattern, whereas at other locations, the angle between the Earth's magnetic (inducing) field and the dip of the body is the factor controlling the symmetry properties of the anomalies. VLF tilt-angle anomalies are similiarly filtered to change a maximum-minimum type anomaly above a vertical conducting sheet into a maximum type of anomaly. This is affected by the Fraser filter, whereas a step further by the Karous-Hjelt filter (1983) goes one more step in the inversion direction by enabling an equivalent current distribution to be produced.

Geophysical inversion is the estimation of probabilities, elimination of impossibilities and less probable choices. Inversion is always iterative, repeatingly self-correcting as long as additional information and understanding of the area/region under study increases. Geophysical inversion is today more than art, it has a definite quantitative touch attached to it. Inversion deals with explaining the real world, nature with its all fabulous richness and variability. Geophysical inversion is a fascinating subject. Let us have a closer look on some of the tools which exist today to assist the geophysicist in this fascinating job.

The development of geophysical inversion techniques has a close connection with advances and development in the world of computers, since inversion always requires the solution to a direct problem, the calculation of the geophysical field of a model. The geophysical community, whether industry dealing with practical aspects of finding resources or basic research institutions looking for models of the structure and composition of the Earth, has always been on the leading edge of computer technology and tried to push the use the computer to its uttermost capacity available. From the interpretation point of view, one can roughly speak of three generations of development in inversion techniques. This is schematically described in Fig. 1.1.

The aim of qualitative interpretation is usually to give an overview of the research area and it has been traditionally been performed by looking at geophysical anomaly maps. Qualitative interpretation is thus often the first operation e.g. in starting prospecting in a new area. In countries, where systematic airborne mapping has been performed, maps for this purpose are available from the respective organization, usually the Geological Surveys. On the other hand, at each stage of a geophysical measuring program, there are stages of qualitative interpretation when each new anomaly profile or data set is first inspected. Advances in workstation technology is quickly computerizing also this stage of operations. Qualitative methods tend to depend more on the experience of the geophysicist than more quantitative methods, even if today their complexity has brought the need of good interpretation expertise into focus again. Some psychological pitfalls of qualitative interpretation will be discussed later.

In quantitative inversion methods (cf. also Table 1.1), the subsurface is described through a model, the details and complexity depending on the geophysical method in use (or actually the ability to calculate the response of the model for that particular geophysical method). The model or models are characterized by model parameters, which attain numerical values to describe the subsurface. Typically model parameters are the size, location, orientation of a body and its physical property (density, susceptibility, resistivity etc.).

6

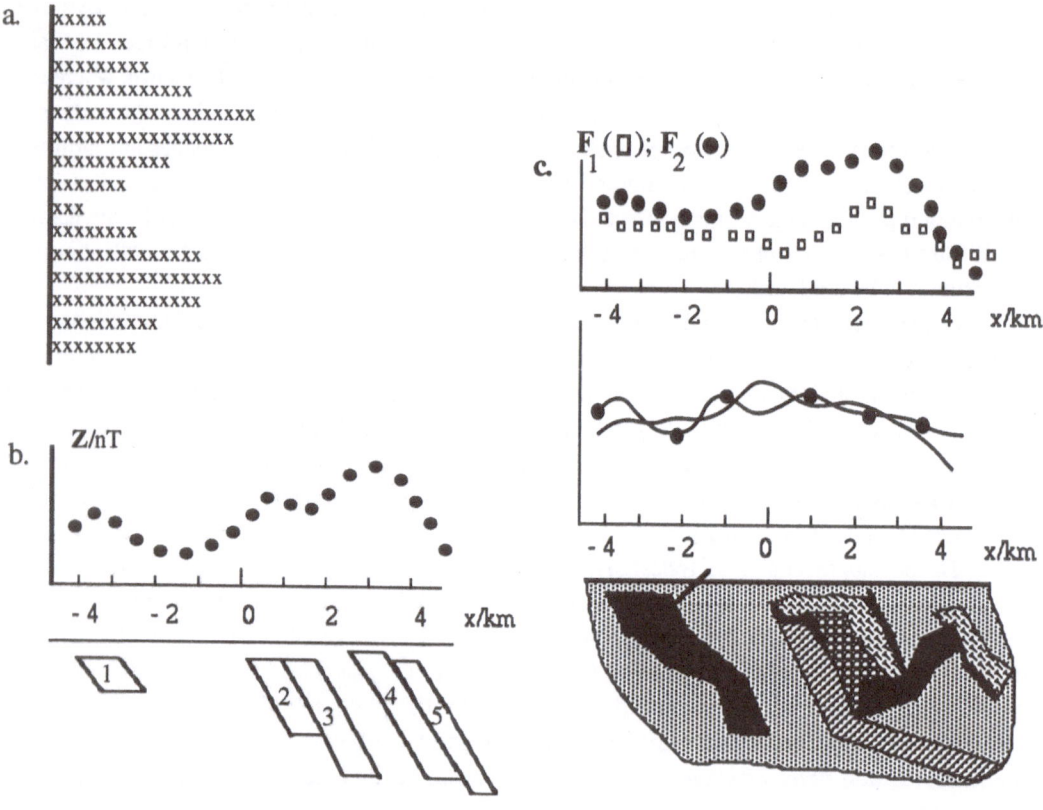

Fig. 1.1 a - c. The generations of interpretation.
a. 1st generation: no interactive operation; "graphical" output only possible on line printer; only one (simple) model body can be modelled at a time; model parameters only on line printer lists.
b. 2nd generation: interactive only for simple models; graphical output on screen possible; only one data profile can be modelled at time; number of models available.
c. 3rd generation: interactive (any action produces a response in less than 3 s); graphical work-station model; several data sets are modelled simultaneously; any part of the model can be pointed at and altered interactively; a great variety of models available.

The simplest method of quantitative interpretation is to select special points or distances of an anomaly profile and connect them via nomograms or "rules of thumb" to parameters of model bodies with simple geometry. This approach was the dominant one before the computer era, but it has its place still today when working in difficult remote places with no or poor access to computing capacity and/or at times of power failure. Scale models were used especially in geoelectrics to provide more information on the properties of various anomaly bodies.

In using special points of anomaly profiles/maps for interpretation uses only a small amount of the information content of the data. Superposition of anomalies, and other forms of disturbances make special point methods useful only for a simple case of one single anomalous body. Much better interpretations can be obtained, when complete anomaly profiles can be used for interpretation. The geophysical field of the model bodies are calculated and compared with the measurements. The model is changed, until a reasonable fit between data and calculations is obtained (cf. Fig. 1.2). Modern computer facilities make this process easily interactive through workstations with excellent graphical properties.

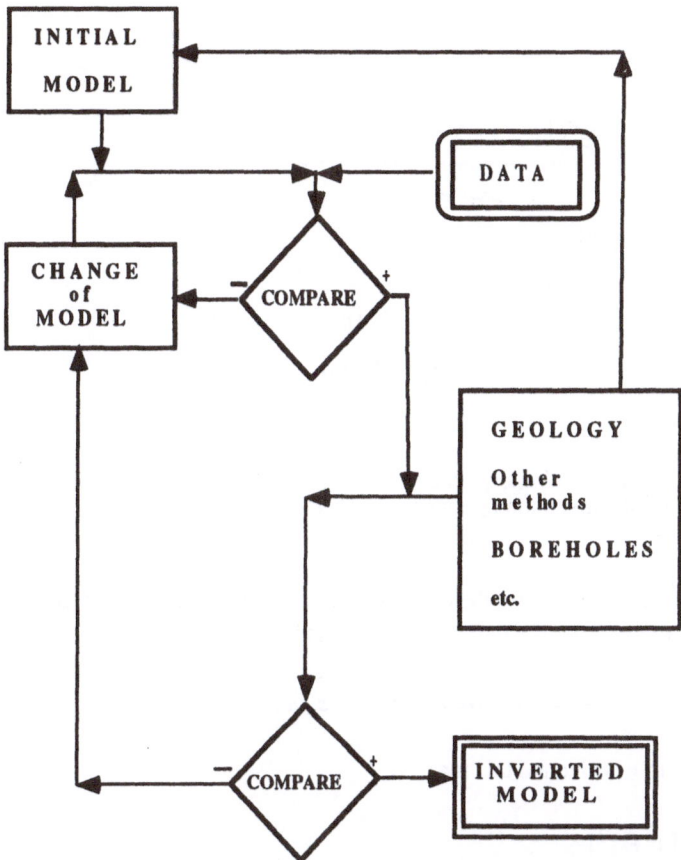

Fig. 1.2. The scheme of iterative geophysical interpretation.

A quantitative comparison of anomaly profiles or maps is possible only if proper measures of fit, mathematically "distance" between functions, can be specified. The sum of squares of the differences between measured and computed anomalies has reached the widest use, but recently other measures have become more popular partly due to their "robustness" and ability to handle the outlier problem more fluently.

The most general group of inversion methods have their theoretical background in information theory and advanced statistical concepts are used. These methods can easily account for correlation within and between various data sets, use of a priori information and other useful concepts.

Any quantitative inversion method can be described by using the concept of *interpretation operator L*. In a broad sense *L* can be defined as a set of rules or mathematical procedures, which, when operated upon the data, will give numerical estimates of parameters describing the model of the subsurface (cf. Fig. 1.3). In an extreme case, in analytical inversion, *L* equals a single mathematical expression. At the other end, for very complex inversion methods, *L* has to be described by a dozen or more steps.

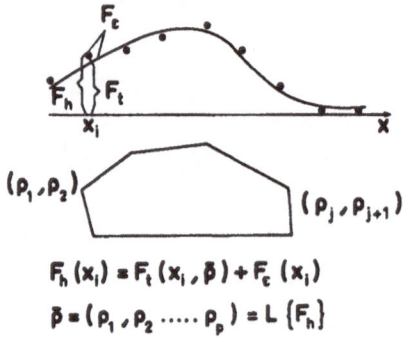

$$F_h(x_i) = F_t(x_i, \beta) + F_\epsilon(x_i)$$
$$\beta = (P_1, P_2 \cdots P_p) = L\{F_h\}$$

Fig. 1.3. Geophysical interpretation. The model, its parameters and field and definition of the inversion operator L. (Hjelt 1973)

In the subsequent examples the concept is described by examples, where the simple basic function of the gravity field of a horizontal line mass is used.

EXAMPLE 1. Inversion operator for least squares inversion

The gravity field of a horizontal line mass is

$$F_{ti} = m / [h^2 + x_i^2] \tag{1.1}$$

We have two parameters describing, the model, its mass/length, m and its depth location. Thus p = (m, h). In the principle of least squares inversion, the function S is minimized

$$S = \sum_{i=1}^{N} [F_{hi} - F_{ti}]^2 = \text{min!} \tag{1.2}$$

leading to

$$\partial S/\partial m = 0$$
$$\partial S/\partial h = 0, \tag{1.3}$$

the solution of which gives numerical values for the unknowns, m and h. After some algebra

$$\Sigma F_{hi}/[h^2+x_i^2] - m \Sigma 1/[h^2+x_i^2]^2 = 0$$

$$\Sigma F_{hi}/[h^2+x_i^2]^2 - m \Sigma 1/[h^2+x_i^2]^3 = 0$$

(1.4)

which can be solved for m directly. The determination of h requires the solution of a non-linear polynomial equation, $P(h) = 0$. The inversion operator is thus:

$$m = \Sigma F_{hi}/[h^2+x_i^2]/ \Sigma\{1/[h^2+x_i^2]^2\}$$

L: $\{$

(1.5)

$$P(h) = 0 \qquad\qquad -> h$$

with

$$P(h) = \Sigma \{F_{hi}/[[h^2+x_i^2]^2\} - \Sigma \{F_{hi}/[h^2+x_i^2]\} * \Sigma\{1/[h^2+x_i^2]^3\}/ \Sigma\{1//[h^2+x_i^2]^2\}$$

Actually different variants of L can be expressed by any of the Eqs. (1.1) - (1.4).

EXAMPLE 2. Analytical inversion

We use the same model

$$F_{ti} = m / [h^2+x_i^2]$$

and model parameter definition $p = (m, h)$. Now we require that at each data point

$$F_{ti} = F_{hi}$$

(1.6)

Defining

$$S_i = F_{hi} / F_{hi+1} = [h^2+x_{i+1}^2]/ [h^2+x_i^2]$$

(1.7)

one can solve the equation easily for h and m is found from any data point

$$m = F_{hi} [h^2+x_i^2]$$

(1.8)

Thus, the inversion operator is:

$$h = \{[x_{i+1}{}^2 - S_i{}^* x_i{}^2]/(S_{i-1})\}^{1/2}$$

L: $\Big\{$ (1.9)

$$m = F_{hi}{}^* [x_{i+1}{}^2 - x_i{}^2] / (S_{i-1})$$

The inversion operator can in most cases be formulated mathematically rigorously as an inverse mapping function. The mapping connects the space of the measurements (data) with the space of the model parameters. For a recent discussion on mapping functions, their properties and use in developing interpretation algorithms, see Oldenburg and Ellis (1991).

1.2   GEOPHYSICAL MODELS

The choice of models has many different aspects. We must be able to compute conveniently and quickly the anomalous field of the selected model bodies. The models must be able to picture

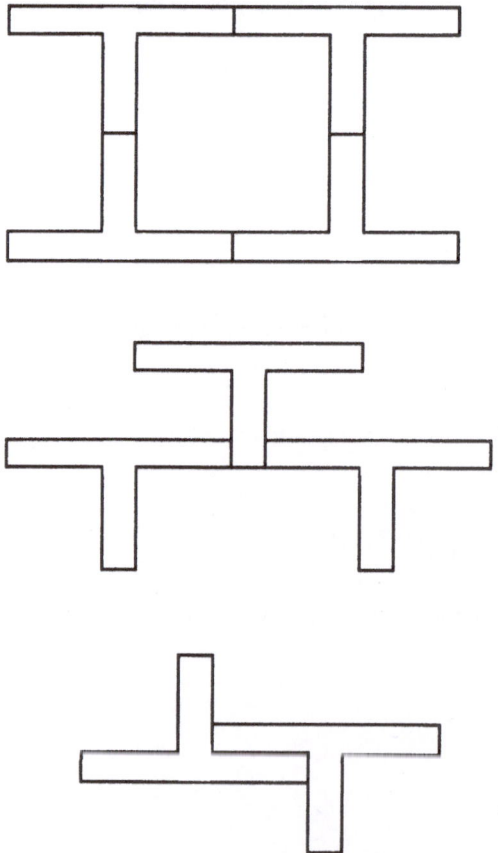

Fig. 1.4. Structures of exposed bedrock are described using one "standardized" building block unit

well a variety of geological situations. The models must not contain too many details either, for which the data do not provide sufficient resolution. In the choice of models, especially for a more permanent computer system, many further details become important. Let us not forget the inherent conservative nature of human beings. This can be nicely demonstrated by a paraphrase of an example given by E. Bono in his book on lateral (today we would perhaps say creative) thinking (1967).

We have found it useful and rewarding to describe structures on bedrock surfaces by a T-model (Fig. 1.4). If the bedrock is partly covered by overburden, the structures are partly hidden. To find them, we have to interpret data, which is partly masked by "errors" (the dark cloud in Fig. 1.5); we tend to use our experience and make a construction of basic T-models, paying attention also to the

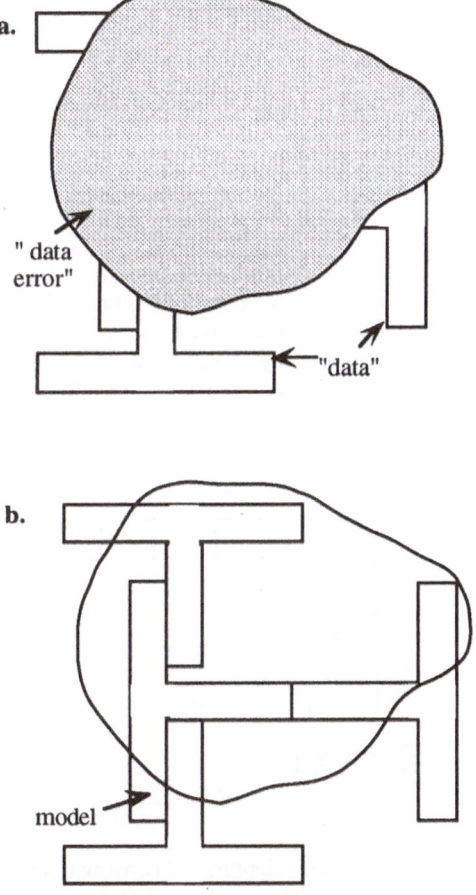

Fig. 1.5 a - b. "Data" obscured by "errors" (**a.**) are "inverted" into a model (**b.**). A model obeying "constraints" (basic units must not overlap; the model has to be "continuous", i.e. no gaps between units are allowed) has been constructed using the basic unit (T).

requirement (a priori information) that the structure behind the cloud is continuous and has no overlapping parts. With the same situation as in Fig. 1.5, we gradually build up an expertise to "see through the clouds", i.e. we become good interpreters - until our model does not work anymore (Fig. 1.6). Then we have to find a new strategy. Depending on how strongly we commit ourselves to the constraints, (continuity and non-overlapping) , on our prejudices and other factors, we may or may not decide to introduce new basic models.

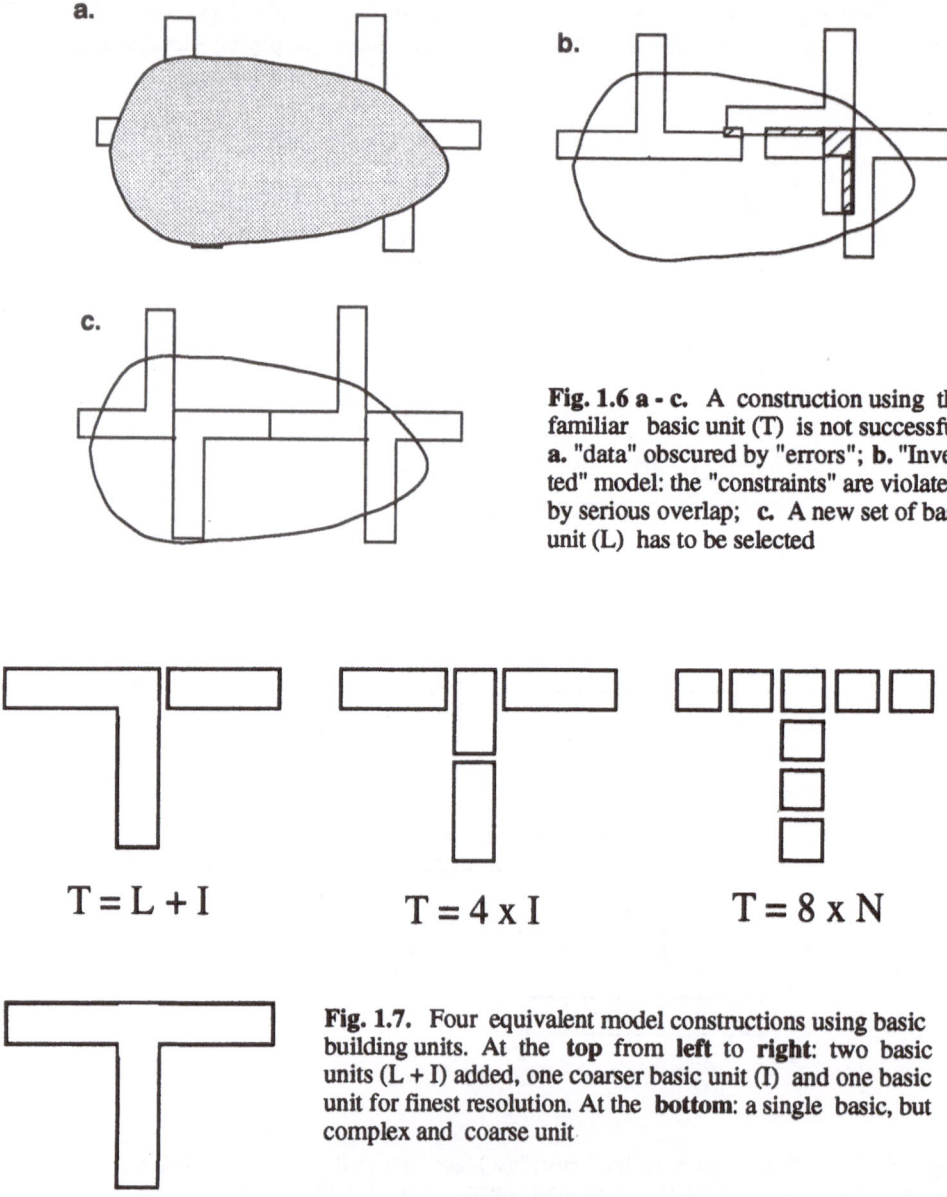

Fig. 1.6 a - c. A construction using the familiar basic unit (T) is not successful. **a.** "data" obscured by "errors"; **b.** "Inverted" model: the "constraints" are violated by serious overlap; **c.** A new set of basic unit (L) has to be selected

$$T = L + I \qquad T = 4 \times I \qquad T = 8 \times N$$

$$T$$

Fig. 1.7. Four equivalent model constructions using basic building units. At the **top** from **left** to **right**: two basic units (L + I) added, one coarser basic unit (I) and one basic unit for finest resolution. At the **bottom**: a single basic, but complex and coarse unit

We soon realize (provided we wish to think creatively) that the original T-model is not such a fundamental model after all. We will have much more flexibility in our modelling, if we make use of two basic models, the L and I, only to realize further, that the L can be constructed out of I's (Fig. 1.7). Or maybe we believe that still more resolution is possible: we finally construct our whole environment out of small squares: the ultimate building block of our modelling universe!

The line of thought followed above translates directly into geophysical modelling (Fig. 1.8). Depending on the wanted accuracy in detail and resolution either crude one-plate or fine-grained block model can be selected.

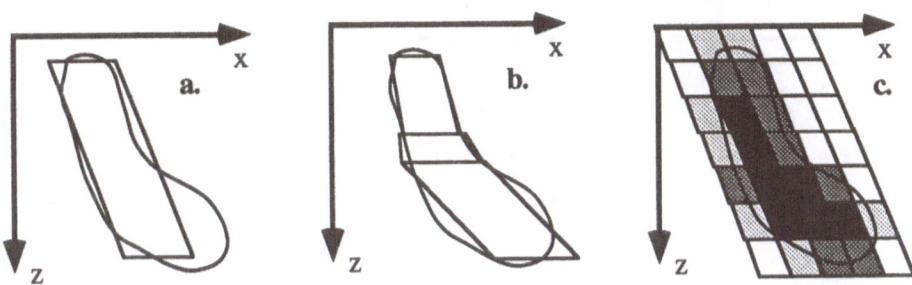

Fig. 1.8. Description of a geological body with various model bodies. **a.** Single 2D plate, **b.** Several 2D plates with varying cross-section, **c.** Block model = many plates with same cross-section

In geophysical modelling the geological description possibilities of models is of great importance, the chosen models should also be flexible to cover many possible modelling situations. Fig. 1.9 shows the three main categories of geophysical models and Fig. 1.10 the flexibility of the dipping prism model. By simply allowing one of the model parameters to approach ∞ geologically quite distinct structures are created. Fig. 1.10 does not exhaust the possibilities of the prism model.

Parker Models

In addition to physically concrete and realistic models, examples of which have been given, various extreme types of models play an important role in inversion, especially in their theoretical and methodological development. Typical extreme models are singular point source, points, lines, dipoles, plane etc. Parker (1980) defines three most important model types for a one-dimensional Earth. Although the definition is made while discussing conductivity models to be used in inversion of electromagnetic induction data, the same defintions apply to the 1-D layered Earth model in connection with any geophysical method.

1. The D^+ model

The model is described by a finite "comb function", the peaks of which are delta functions:

$$\sigma(z) = \sum_{i=1}^{N} b_i \, \delta(z - z_i)$$

The model has an equivalence in circuit theory: a chain circuit, where the forward branches consist of resistors and the perpendicular branches of capacitances.

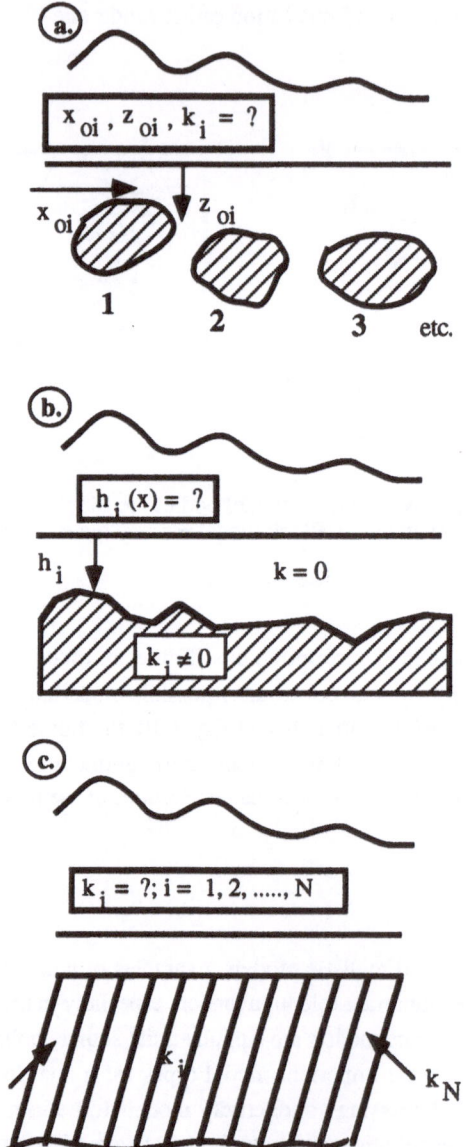

Fig. 1.9. Three principal types of geophysical models. **a.** Separate bodies; **b.** boundary between two rock types; **c.**block model with one basic model body

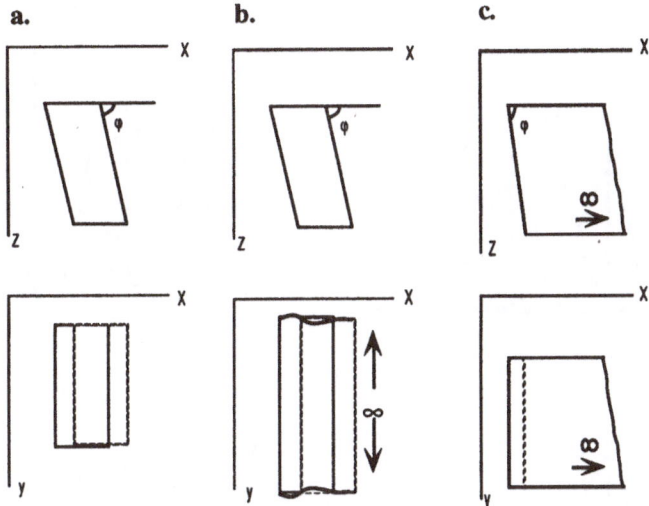

Fig. 1.10. Vertical and horizontal projections of the dipping prism model and its extensions.
a. The prism model; **b.** its 2D equivalent, the thick plate; **c.** further extension to a horizontal step

2. *The H+ model*

The model consist of a stack of layers, where for each layer k the parameter $\sigma_k h_k^2 =$ constant. In geoelectric methods σ is the layer conductivity and h the layer thickness.

3. *The C2+ model*

The model is constructed from functions which are
◊ strictly positive
◊ twice differentiable
The model has been used by Weidelt (1985) in his studies of extremal models for electromagnetic induction.

The resolution of inversion will naturally also be affected by the choice of model. One naturally should prefer a basic model with the sharpest unmerged anomaly. This is illustrated by the simple naive example of Fig. 1.11. Assume that two bodies, each producing a maximum-type anomaly, are located close to each other. As long as the merged anomaly has two distinct peaks, no mistake can be made: two model bodies are required to model the data. Even for strong merging, the increased width of the anomaly awakes our suspicions. If the anomaly of the basic model is of the minimum-maximum (or vice versa) type, it is broader and the neighbouring anomaly parts have a tendency to cancel each other. The result is "false" zero-crossing and other non-indicative properties of the merged anomalies (Fig. 1.11).

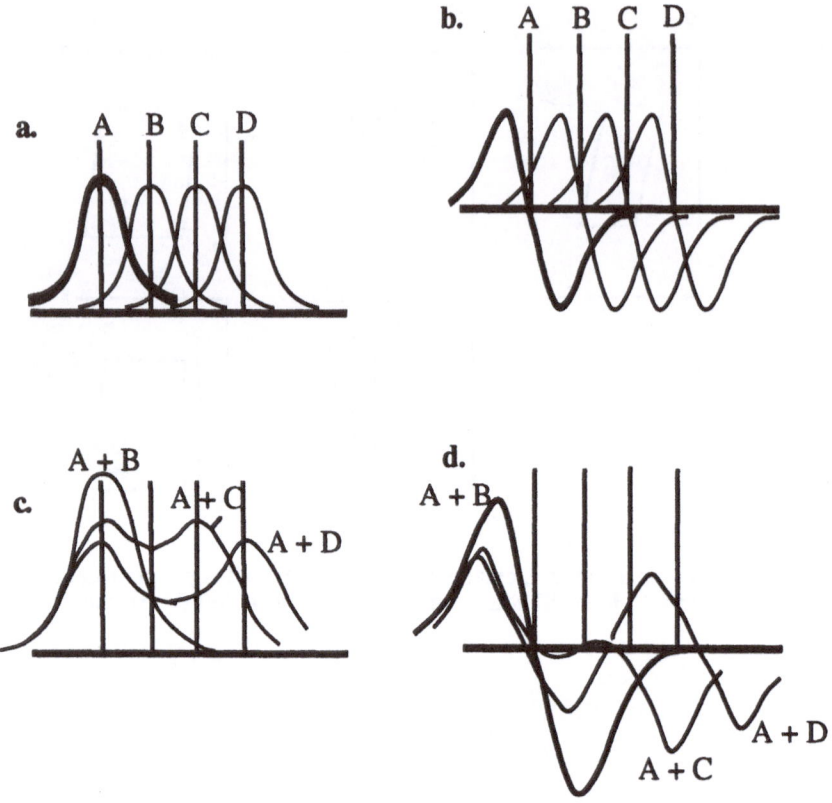

Fig. 1.11. Two basic forms of anomaly profiles and the influence of their horizontal extension on the sum anomaly. The models, causing the basic anomalies are assumed to be located at A, B, C and D (at the vertical lines). The sum anomalies are denoted with A + B, A + C and A + D, respectively.
a. Maximum type basic anomalies; **b.** Maximum-minimum basic anomalies; **c.** Sums of maximum-type anomalies; **d.** Sums of maximum-minimum types

1.3   MODEL PARAMETERS

The question of model parameters will be amply described from various points of view within the chapters of this book. Choice of non-correlating parameters, choice of relevant parameters (from the point of view of model description), linearity versus non-linearity of the parameters and keeping the amount of (different) parameters at minimum are some of the important topics that will be discussed further.

1.4 FITTING MODEL FIELDS TO DATA

The human eye is an excellent correlating device. Therefore the visual comparison of several theoretical data sets, profiles or maps with the measured data sets quickly tells the interpreter which of the model fields fits the data best. In order to do the same automatically (or semiautomatically) on a computer, one needs a quantitative *measure of fit* between the data and model fields. An interpretation algorithm (method) then consists of a set of rules which defines how to minimize this measure.

Mathematically the measure of fit is equivalent to defining a distance between the sets or functions of data and model fields. Since measurements normally are taken at discrete points of a profile or on an area, the functions are not continuous but discrete. If e_i is the difference between the measurement $F_{hi} = F_h(x_i)$ and the model (theoretical) field $F_{ti} = F_t(x_i)$, then the most commonly used measures of fit are (cf. also Fig. 1.12):

The sum of squares	$S_1 = \sum e_i^2$
The weighted sum of squares	$S_2 = \sum w_i\, e_i^2$
Maximum difference	$S_3 = \max |e_i|$
Correlation coefficient	$S_4 = r = \dfrac{\sum\limits_i F_t(x_i)\cdot F_h(F_i)}{\sqrt{\sum\limits_i F^2{}_t(x_i)\cdot \sum\limits_i F^2{}_h(x_i)}}$

Minimizing S_1 or S_2 leads to the well-known method of least squares (LSQ) , minimizing S_3 defines the Minimax method. The theoretical background of LSQ is well developed, so that it is natural that the majority of geophysical inversion methods are formulated using this criterion. The method has a firm statistical background and works well, when the errors of the fit between measurements and the theoretical models, e_i, have a normal (Gaussian) statistical distribution (cf. Chap. 5).

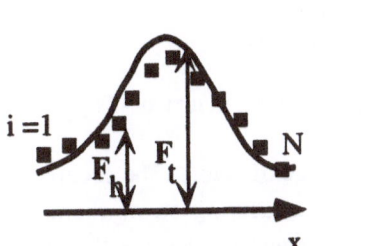 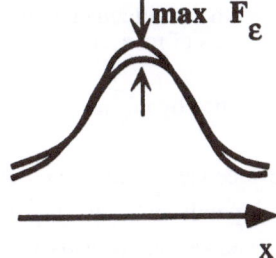

Fig. 1.12. Measures of fit between data and model field profile. At **left**: discrete data F_{hi}, continuous model field F_t; at **right**: continuous data and model field, maximum deviation

The question of the statistics of geophysical inversion errors has been addressed very little in the literature. As a matter of fact, recently several authors have questioned this approach and introduced "robust methods". In discussing the basics of such methods Steiner (1988) by the M-fitting (adjustment according to the most frequent values), states that robust methods demand significantly more computing time than the classical LSQ technique. However, since the costs of data acquisition are high, the efficiency of data processing and interpretation is worthwhile. Robust methods also provide a natural and efficient framework for handling the problem of data outliers (cf. Chaps. 3 and 4)

1.5 FACTORS AFFECTING INVERSION

In geophysical inversion, especially when designing computer programs or program systems certain aspects deserve special attention, for example:

Measuring errors and accuracy

Inaccuracy (incompleteness) of models	A	Fig. 1.13
Superposition of anomalies	B	Fig. 1.14
Coupling between model parameters	C	
Non-uniqueness of geophysical inversion	D	Figs. 1.15 and 1.16
Insufficient data point density	E	Fig. 1.17
Equivalence	F	Figs. 1.18 - - 1.20
Special features of various methods	G	

e.g.: * demagnetization
* geometric attenuation of EM source fields
* effects of measuring geometry (DC and EM)

Each of them are described by means of a simple example (A - G in text and Figs. 1.13 - 1.20)

A. A model of the subsurface can never be an exact description of nature. The accuracy and the resolution of inversion depends both on the amplitude of measuring errors themselves, the geophysical method used, and also on the capacity of the chosen model to describe the subsurface in sufficient details. Therefore each chosen model - often a body of simple geometrical shape - and its parameters are determined inaccurately. Schematically this is shown in Fig. 1.13. Instead of the exact anomaly shown by a continuous line, the data values vary inside the hatched area. If the exact curve and the envelopes of the regions of data variation are inverted, we get three models and actually we can only state, that the case of the measured anomaly is somewhere within the crosshatched area. The figure is exaggerated, but an obvious result is that the details of the models closer to the measuring profile/plane can be interpreted with greater resolution than more distant details of the model.

B. If a measuring profile, a ground magnetic profile - say, contains several closely spaced anomalies, they merge more or less together, and a superposition of anomalies takes place. In Fig. 1.14 to the left the horizontal resolution is sufficient to distinguish between two separate model bodies, to the right the situation is unclear because of model errors, if the error regions of the models partly overlap. See also Fig. 1.11 where a more concrete example related to the form of the anomaly was discussed.

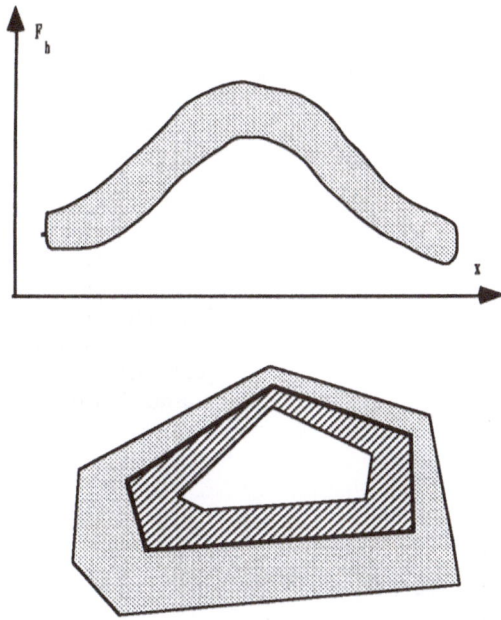

Fig. 1.13. Schematic effect of data errors on the accuracy of the subsurface model

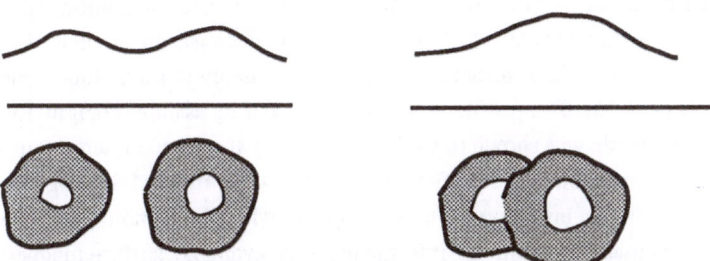

Fig. 1.14. Spacing of anomalies and model resolution (schematically). **Left**: Anomalies are far enough for the model bodies to be distinguished; **right**: overlapping of anomalies results in poor resolution of the model bodies

C. Some model parameters affect the theoretical anomaly in a similar way. If this correlation between the parameters is high enough, the effect of the parameters can not be separated during inversion, but only a combination of the parameters can be resolved in inversion. The classical example is the determination of size and density of a spherical or horizontal cylinder in gravimetry. Since the total excess mass is the factor affecting the anomaly, the density contrast and size cannot be determined separately unless further information on either of these parameter is available. Thickness and depth to the top of a thick 2-D plate-like model in magnetometry and the S-value (conductivity x thickness product) in most EM methods are further examples of strong correlation between parameters.

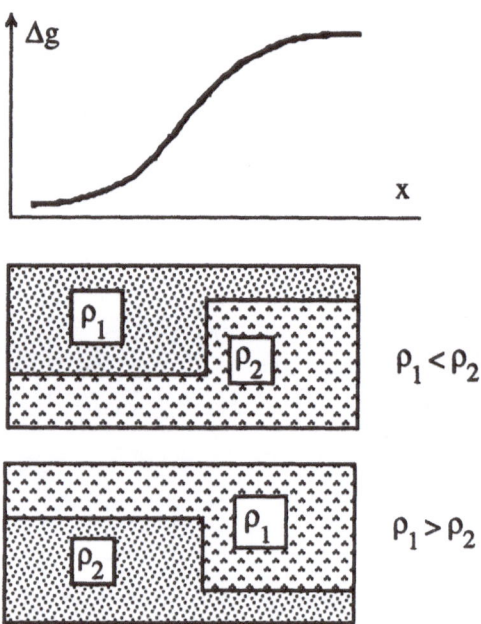

Fig. 1.15. On the non-uniqueness of a gravity anomaly. The two models cannot be distinguished without further information, e.g. geological feasibility (= the denser material lies below). (Redrawn from Parasnis, 1975)

D. Most geophysical fields, but notably potential fields (gravimetry or magnetometry) are nonunique in the sense that there exists an infinite number of models, which may explain the data within the limits given by the accuracy of measurement (Fig. 1.15). All additional information, inversion of other geophysical data, geological data etc. have to be used to decrease the amount of non-uniqueness. One example is shown where the aeromagnetic total field anomaly of a synclinal structure was computed theoretically (Fig. 1.16). The profile is directed N-S and its assumed height 150 m above ground.

When this profile was shown to several groups of experienced interpreters, they systematically started a 2-D interpretation of the profile by selecting two northward dipping plates. A similar result was obtained by using an inversion computer program, where both anomaly selection and the iterative inversion was automatic. If additional information were available: surface magnetometry, gravimetric data or a geological hypothesis, inversion would more probably lead to the synclinal model. Already aeromagnetic data taken at 30 m height above ground (not shown) contains information on the horizontal part of the syncline. The gravimetric profile, although lacking the minor details of a ground magnetic profile, reflects clearly the symmetry of the synclinal structure. The example clearly demonstrates the loss of detail, when data are taken farther away from the source structure.

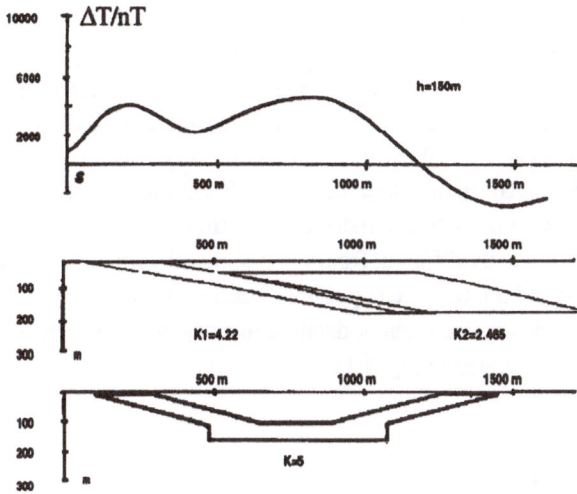

Fig. 1.16. Two magnetic 2-D models producing the same total field anomaly profile at a height 150 m above the ground surface. The anomaly profiles at lower heights and at the ground are different from each other. Without additional information the two models cannot be distinguished.

E. It is easy to understand that insufficient data point density directly may affect the resolution of inversion. This may be caused by lack of resources, either in time or money available or just by improper experimental design. An example, where the economic aspect of the problem is nicely demonstrated is presented by Parasnis (1973). The example is an unorthodox use of magnetic profiling.

The problem was to determine the depth to the bedrock in a region covered by Quaternary material. The bedrock was of interest for prospecting, since some boulders containing interesting amount of mineralization had been spotted on the surface. For further studies of the bedrock and to take samples, it was necessary to remove the Quaternary material. If the thickness of the surface material were smaller than 3 m, it would be cheaper to remove the material to study the bedrock surface directly over a larger area. For greater depths, sample drilling would be the cheapest way to study the mineralization of the bedrock.

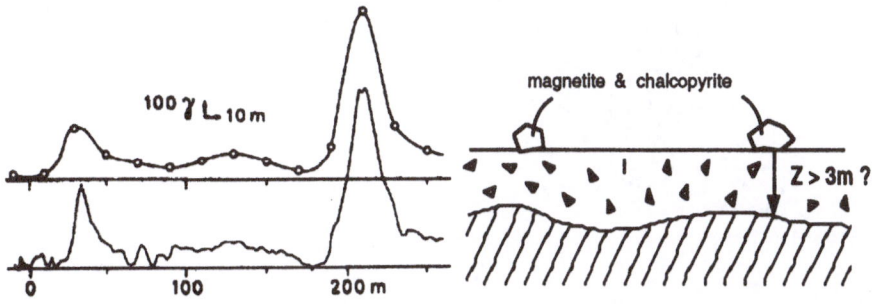

Fig. 1.17. The influence of point density on determination of thickness of overburden from magnetic profiling. The bedrock is assumed to contain magnetic minerals. The profiles to the **left** were measured with 20 m (**upper**) and 2 m **(lower)** point separation. The geological problem is described to the **right.** (After Parasnis 1974)

Since the boulders were magnetic its was assumed that the mineralized parts of the bedrock would show a magnetic anomaly. To determine the depth to the bedrock surface by magnetometry it was necessary in addition to assume that the mineralization extends to the surface. A magnetometer profile with 20 m point density existed in the area. Inverted, some clear anomalies indicated a depth to the bedrock surface in the order of 13 m. A special profile with 2 m point density was measured, indicating that the main anomalies in fact were much sharper, their cause thus being more close to the surface. A re-interpration of the new profile gave 5 m to the bedrock surface, still not quite sufficient for using a cheaper removing method, but significantly different from the first interperation. Without the increased point density this information would not have been possible. Each geophysical method has its characteristic point density in relation to depth resolution or depth of investigation. Insufficient profile length can also be problematic as is demonstrated in Fig. 1.18.

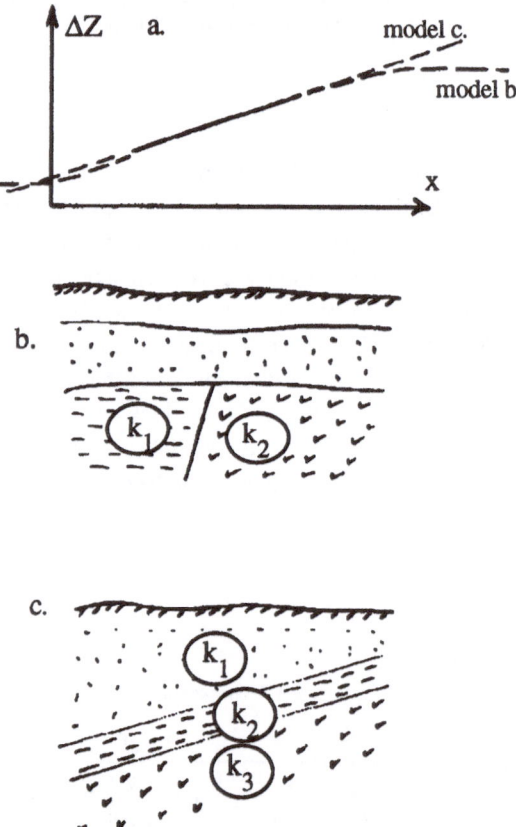

Fig. 1.18 a - c. The influence of profile length on magnetic inversion. **a.** If the anomaly profile shown by the full line is only used, the models **b** and **c** cannot be distinguished in inversion. **b** and **c.** Two geologically completely different models of the subsurface, which produce the same anomaly along the part drawn in full.

F and **G.** Most sounding methods suffer from equivalence. The classical example is DC sounding. Any part of the subsurface containing magnetic material is magnetized in the direction of the existing geomagnetic field. The vectorial field of these magnetized structures have very clear directional properties, allowing the determination of the size and location of the structures. Magnetometric data is suitable for determining dips of elongated 2D structures. For very strong magnetization, however, demagnetization effects tend to change the direction of magnetization, thus distorting e.g. dip determinations. The remanent magnetization, caused by ancient fields with differing directions of the Earth's magnetic field with respect to the formations, is specially harmful for elongated, moderately magnetized sulphide rocks. Demagnetization effects in interpretation can rather easily be corrected for numerically, but remanence correction is possible only if petrophysical information from directional samples in the region is available.

The electrical equivalence and the hidden layer problem is typical for both electrical and electromagnetic sounding. Their counterparts exist also in seismic sounding. Equivalence can be considered as a case of parameter coupling. When the conductivity and thickness of a layer varies within certain limits (their product remaining constant) no differences can be seen in the sounding curves within the accuracy of measurements. The classical example has been presented by Homilius (1970; Fig. 1.19), where the basic model contains a middle conducting layer with a conductance of 20 S. If this layer is replaced by three thinner layers, but with the same total conductance (Table 1.2. and Fig. 1.19), then all sounding curves are within ± 5% of the curve for the basic model. Since the Earth is layered (1D model), the true model instead of an equivalent layer can be obtained only by using additional information, e.g. drill hole data. A change of electrode geometry does not provide a solution to the equivalence problem if the Earth is truly 1D. One can also speak of a more general

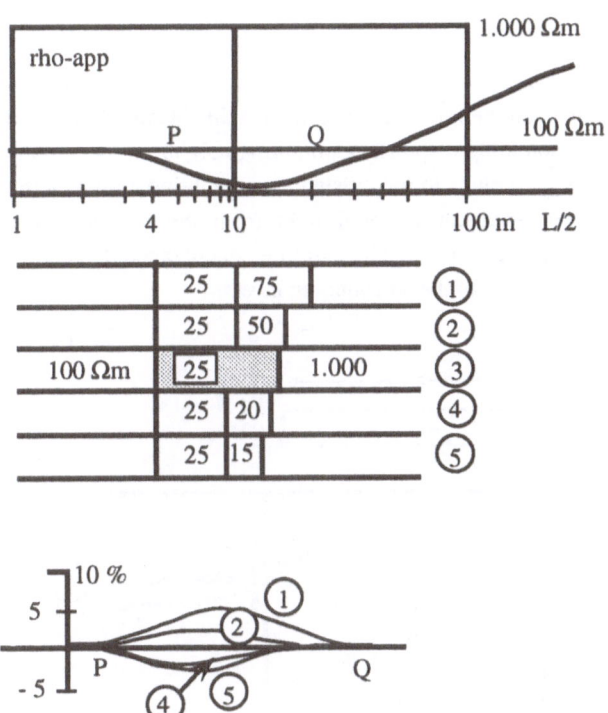

Fig. 1.19. The "classical" example of equivalence in DC sounding. Five alternative 1D (layered Earth) models produce the same (within data accuracy of 5 %) apparent resistivity sounding curve. The S-values of the models are given in Table 1.2. (Redrawn from Homilius 1970)

equivalence (Fig. 1.20), when three completely different structures (but similar conductivities) produce similar sounding curves. Since the structures of Fig. 1.20 b, c are 2D, taking a sounding at a
different point will change the sounding curve, thus allowing the equivalence problem to be solved.

TABLE 1.2. Table of S-values for the models of the DC equivalence (Fig. 1.19)

$$S_1 = \frac{6}{25} + \frac{9}{75} = \frac{9}{25}$$

$$S_2 = \frac{6}{25} + \frac{6}{50} = \frac{9}{25}$$

$$S_3 = \frac{9}{25} \qquad\qquad \text{the "basic" model}$$

$$S_4 = \frac{4}{25} + \frac{4}{20} = \frac{9}{25}$$

$$S_5 = \frac{4}{25} + \frac{3}{15} = \frac{9}{25}$$

In the hidden layer case, a conducting layer sandwiched between two less conducting layers is
too thin to give rise to sufficient geometrical damping (in the DC case) or amplitude absorption (in
the EM case) to affect the sounding response. Comparing neighbouring soundings along a profile,
borehole information may give clues to the existence and continuation of a thin conducting layer.
Sometimes only indirect evidence can be stated in the form: the measurements do not exclude the
possibility of a (hidden) conductive layer. The extreme conductance (S-value) can be theoretically
assessed for known conductivities of the surrounding layers.

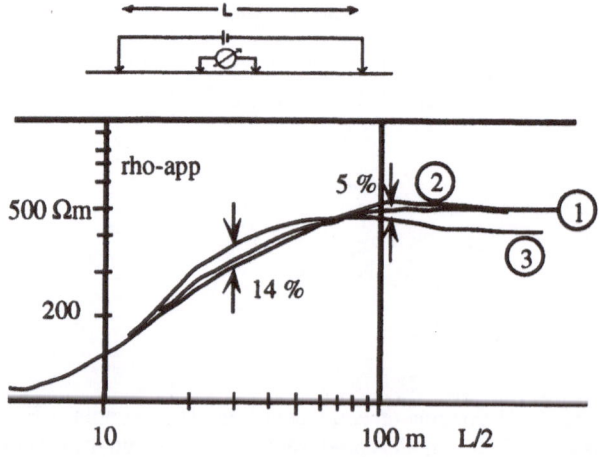

(Fig. 1.20 continues)

(Fig. 1.20 continued)

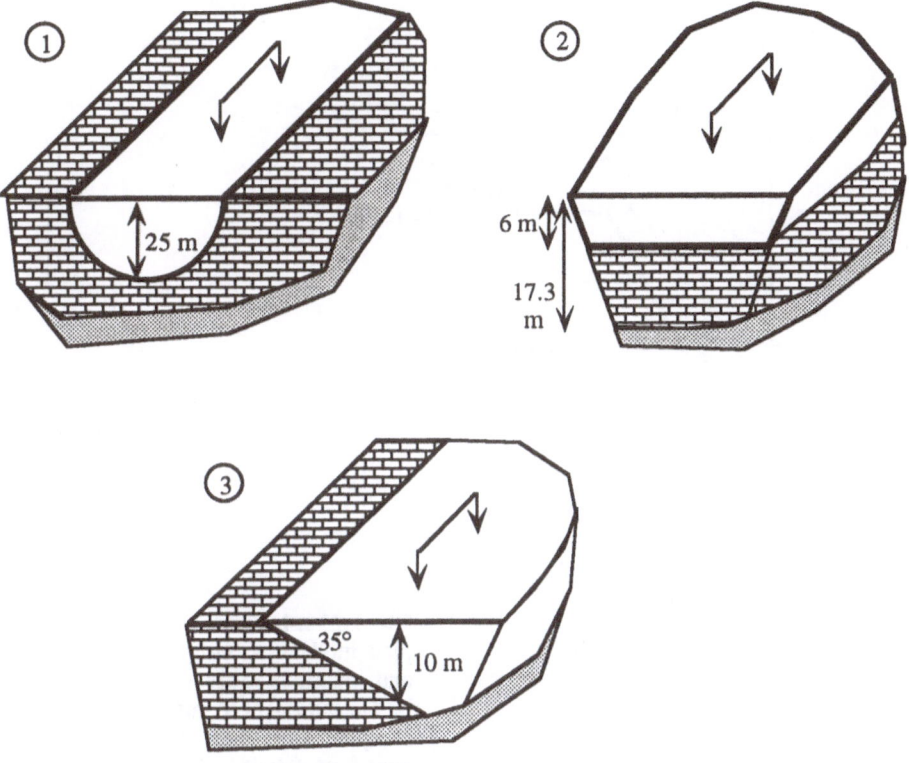

Fig. 1.20. The 2D equivalence of DC sounding. Models **1-3** produce apparent resistivity curves, which differ from the average less than ±7%. The equivalence does not hold, if additional sounding points are considered. (Redrawn from Homilius 1970)

1.6 COMPUTATIONAL ASPECTS

For quite some years, the limitations of the calculation capacity of computers was the limiting factor both in the direct problem, the calculation of geophysical fields and anomalies of given bodies or the inversion part proper. Many people seem to believe that there is no longer any need for consideration on the calculational aspects. However, with the increasing capacity of microcomputer, desktop or laptop, portable or not, in the field or at home, the demand seems to have been increasing at least with the same speed as the technical capacity of computing devices. More complicated models are needed, the volume of the data sets is continuously increasing and more and more concern is given to the fact, that the costliness of data acquisition also deserves the best of data processing and inversion methods to get the most information out of the data. Also the newer developments in inversion techniques have increased the possibilities of designing measuring campaigns and layouts more adequately than before.

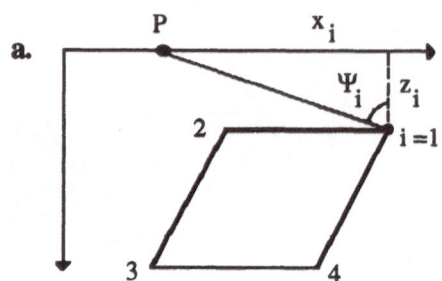

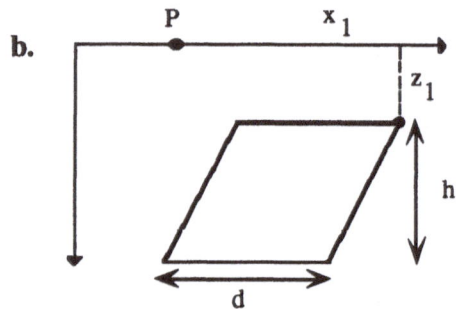

Fig. 1.21 a-b. The 2D magnetic thick plate model. **a.** Original parameters and equations as given by Gay (1963). **b.** Definition of parameters for the computer-time optimized version of equations. (Hjelt 1973, 1975)

Fig. 1.21 shows one of the most popular two-dimensional (2D) models of magnetic interpretation in the 1960s and 1970s, the magnetized thick plate. Gay (1963) collected some of the features of the expression for the magnetic field of the plate into a compact expression and useful set of characteristic curves. He minimized the number of descriptive parameters and any of the field components could be determined from the same set of curves. These expressions became the basis for many interpretation systems, including one early desktop implementation by Hattula (1970). The equation is (for notations, cf. Fig. 1.21):

$$F(P) = b \cdot \{ \psi_1 - \psi_2 + \psi_3 - \psi_4) \cdot \sin \varphi + \ln \left[\frac{\cos \psi_2 \cdot \cos \psi_3}{\cos \psi_1 \cdot \cos \psi_4} \right] \cdot \cos \varphi \}$$

with $\psi_i = \operatorname{atan} [(x_i - x_P)/z_i]$ and φ being the dip of the model. This system was very slow and the production of an anomaly profile (of a single plate or very few plates) could take minutes. When modifying the system for a central computer the demand for more complex models increased and that together with the introduction of non-linear optimization codes caused computing time to become a system bottle neck (Hjelt 1973, 1975).

A deeper analysis of the thick plate equations showed that considerable saving in computing time was possible. Two factors were crucial and they are still valid today, especially when trying to gain the greatest performance of the supercomputers or massively parallel computers. These are identifying the most time-consuming parts of the expressions and eliminating every unnecessary repetition of calculations (whether time-consuming or not). The costliest (in time) computer operations are the special functions, *trig, log, atan, sqrt.* By rearranging the equations of the thick plate, using the geometry of the problem, the addition rules of logarithmic and trigonometric functions and by separating those calculations which need to be done only once for a profile from those necessary at every point of the profile, a four to five fold increase in computation speed was reached (cf. Table 1.3). The optimal formulation is :

$$F(P) = b \cdot atan\ Q + c \cdot \ln R$$

where Q and R are polynomials of d, h, x_1 and z_1 (Hjelt 1973, 1975). Some additional book-keeping procedures are necessary in the new formulation to take care of the multivalued properties of trignometric functions (the *atans* are especially tedious). Although the amount of increase gained will differ from one computer to another, the principle remains the same and such time-cost reducing procedures really are worthwhile for any geophysical program, which will be used more systematically in practice.

TABLE 1.3   Number of computer operations needed to calculate the magnetic field of a thick 2D plate

	±	*	/	*atan*	*trig*	*ln*
Direct	8N -1	4N	4N	4N	4N	N
Optimal	13N -1	12N	4N	N	0	N

N = number of data points in a profile

Another example of how proper arrangement of the computing operations will save computer time is indicated in Figs. 1.22 (2D case) and 1.23 (3D case). When a block model of the subsurface is used, it most often consists of repeating the field calculations of same basic model. Consider Fig-1.22., where the block is constructed out of M (similar) vertical thick plates. A straightforward calculation of the anomaly profile would require the field of a plate to be determined at M*N anomaly points. However, thinking over the situation, one soon realizes, that the anomaly of plate i (say) at point j (say) is exactly the same as the anomaly of plate i + 5 at point j + 5, that of plate i - 3 at point j - 3 and so on, with the exception of a linear multiplicative factor, the physical property of the plate, which varies from plate to plate. The physical parameter can be the amplitude of magnetization. susceptibility, density or whatever property is necessary for the field one wishes to calculate. In summary, it is sufficient to calculate the anomaly of unit physical property 2M times, store the results and produce the profile by simple multiplications.

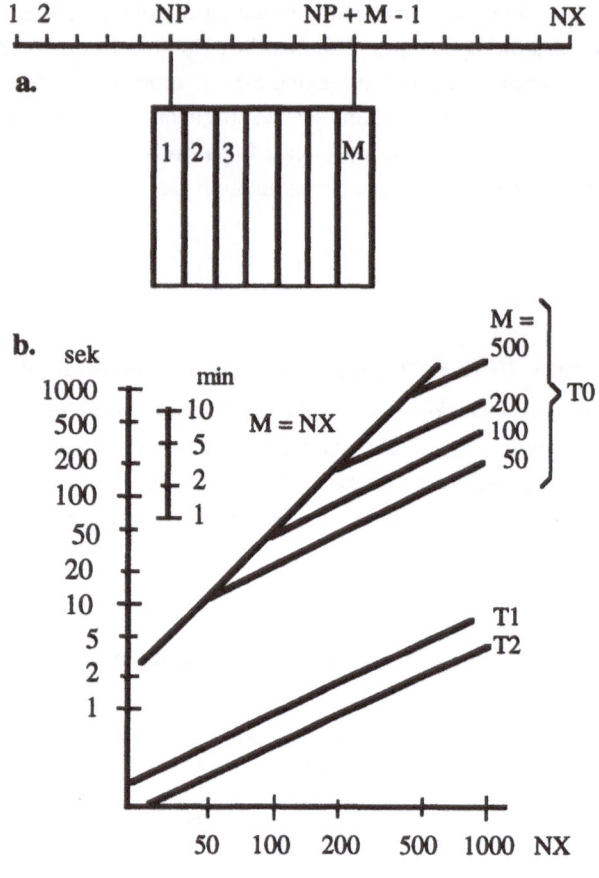

Fig. 1.22a. The 2D block model consists of M rectangular thick plates. The anomaly has to be computed at NX field points. **b.** Computing times for the magnetic 2D anomaly profile with NX field points for various numbers of plates, M, in the block. The anomaly of a plate is computed

TO: M x NX times,
T1: 2 x M times,
T2: M times.

The time scale is logarithmic

The result of such an analysis is depicted in Fig. 1.22. The number on the vertical axis are actually irrelevant, although they are valid for a specific real computer system (IBM 360/40 as far as I remember). The principle just explained gives a gain in computer speed of 2 decades (2 orders of magnitude) or more. Carrying out a similar analysis for the 3D case, where the anomaly is to be calculated over a research area, at NX* NY points (Fig. 1.23), gives a similar result. The basic anomaly of a 3D body (vertical prism, e.g.) for unit physical parameter needs to calculated only at 4*MX*MY points and the gain in using this principle over straightforward calculation depends on the size of the research area, but is on the average 2 decades also. In addition to time considerations all techniques of good computer programming practice deserve proper attention in designing geophysical computer systems, whether for anomaly calculation only or also for inversion.

One of the most fascinating developments in the computer science has been the growth of parallel computation. Since a considerable part of the human "data processor", the brain, works in parallel mode and many subtasks of the complete cycle from producing geophysical data to constructing a model of the subsurface are inherently parallel, the promises of truly multiprocessor computing systems (the most massive ones having more than 60.000 processors today) are great indeed. The developments in parallelism will probably parallel also the most promising trends and developments in inversion methodology. Some very elementary considerations on these new aspects in geophysical interpretation can be found in Chapter 9.

a.

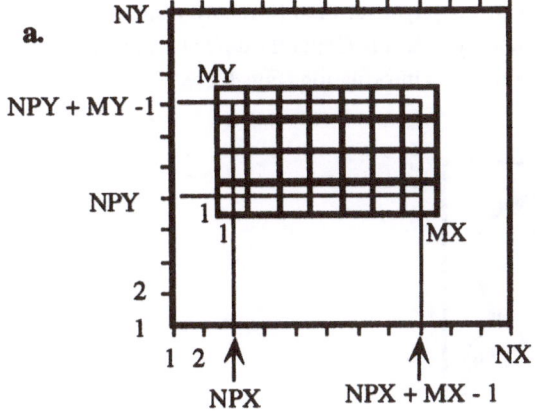

Fig. 1.23a. The 3D block model consists of $MX \times MY$ rectangular vertical prisms. The anomaly has to be computed at $NX \times NY$ field points. **b.** Computing times for a magnetic 3D anomaly at $NX \times NY$ field points. The anomaly of a plate is computed

TO: $M \times NX \times NY$ times,
T1: $4 \times M$ times,
T2: $2 \times M$ times.
$M = MX \times MY$ in the block. The time scale is logarithmic.

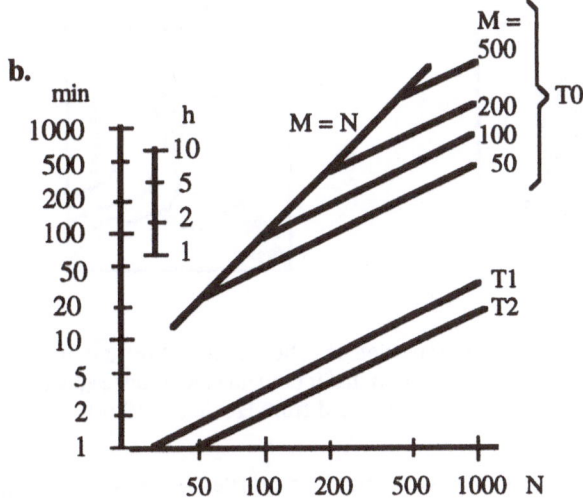

1.7 PSYCHOLOGICAL ASPECTS OF INVERSION

Especially in qualitative inspection of data, but also at all other stages of geophysical inversion, one should be aware of some pitfalls, which both has and has not to do with the nature of geophysical measurements. Geophysical data is information which can be interpreted in various ways. The inversion problem is said to be nonunique. Several, in cases of high noise level even widely differing, models can explain the data within the measuring error. The interpreter therefore has to make choices between various possibilities. Without enough other data or facts, it is human to restrict oneself to model alternatives, to which he/she is used. If the geophysicist has mainly worked in regions, where the structures and geological units have vertical boundaries, he may have difficulties in recognizing the anomalies of horizontally or subhorizontally dipping bodies (cf. the case of the magnetized syncline in Fig. 1.16). It is also considerably easier to "see clearly" structures on an anomaly map, which support the preconceptions of a geologist or geophysicist than to identify structures in an unknown area. The psychological nature of this conservatism is elegantly described in many textbooks on psychology by pictures which contain ambiguous information (Fig. 1.24).

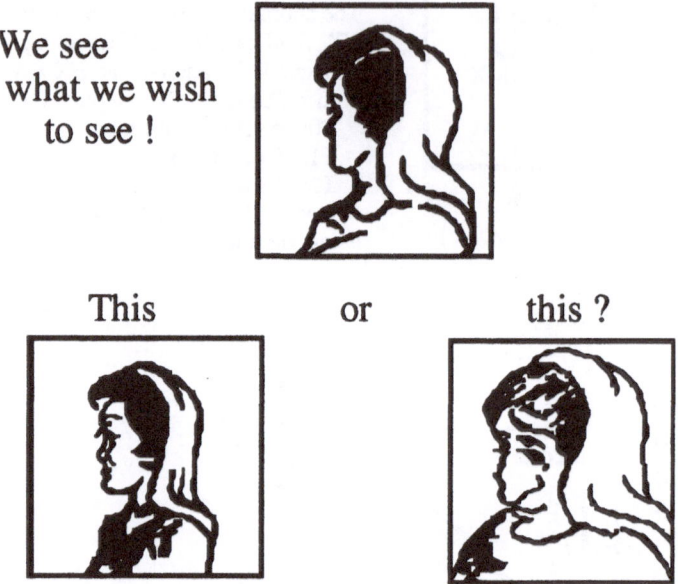

We see
what we wish
to see !

This   or   this ?

Fig. 1.24. The psychological dimension of pattern recognition and non-uniqueness. We tend to interpret non-unique data so that it coincides with our expectances, experience and perhaps also prejudices. (Adapted from Hochberg 1964)

In addition to ambiguity, geophysical maps and profiles always contain disturbing noise. How well certain features (e.g. long uninterrupted fracture zones) can be recognized on a geophysical map, depends on the nature, distribution and amplitude of the disturbing fields. If the noise field has same directional characteristics as the signal we wish to identify, it may be difficult or even impossible to make a reasonable qualitative interpretation. A simple example from outside geophysics, again from psychology, is given in Fig. 1.25.

Pattern recognition has been studied widely both for the purpose of identifying objects from (mainly digital) pictures and for the purpose of enhancing noise-contaminated picture (e.g. in remote sensing applications). Although certain advances have been noticed over the years, still automatic pattern search is still one of the difficult subjects of computer science. The human eye after all is one of the best correlators, but the habits and conservatism of the human brain and mind often plays games even with experienced people. The famous case of the channels on planet Mars is a good example to this effect. Many refreshing thoughts can be found in a recent article by Raiche (1991). If one wishes to go deeper into the philosophy of inversion, pattern recognition, whatever you prefer to call interpretation of measurements, one has to consult such books as Moser (1987), or Popper (1986, revised edition), to mention just two.

Fig. 1.25. "Noisy" data from a general point of view. The number **4** and the letter **H** represent error- free "data" and all additional lines "noise". The important pattern (the error-free part of the data) can be easy or difficult to recognize, depending on how "similar" the pattern of noise is. (Adapted from Hochberg 1964)

1.8 WHAT THEN IS PRAGMATIC INVERSION?

In this book the approach has been to go through and explain the most common techniques used in inversion of data. The emphasis is less on mathematical elegance and completeness (there are a plenty of excellent advanced textbooks for that purpose) but more on the description of how these techniques work, how they have been applied in geophysical surveys using a variety of methods, what are the limitations and pitfalls. The examples have been chosen so that they will aid improvements and constructive rethinking and revaluation of our habits as geophysical inverters. That is what I have called pragmatic inversion of geophysical data.

Whether you are involved in inversion of real geophysical data or designing a system to assist in that task or experimenting with your favourite "idea-of-the-century" algorithm, the black box which is going to solve all the mysteries and dilemmas of geophysical interperation for now and forever, please give ten minutes of your time to the following questions:

1. What are we looking for ?

2. What information is available ?
    ◊ can we design our experiment or do we have instrumental limitations
    ◊ geological knowledge
    ◊ borehole data

3. Are our models relevant ?
    ◊ how to handle non-uniqueness

4. How much sophistication can we afford ?

These are some of the questions that are approached in this book and I hope to your benefit!

REFERENCES

de Bono E (1967) The use of lateral thinking. (Finnish translation) WSOY.

Gay SP Jr (1963) Standard curves for interpretation of magnetic anomalies over long tabular bodies. Geophysics 28: 161 - 200

Hattula A (1970) On the regional interpretation of geophysical measurements. M.Sc.Thesis. Helsinki Univ Technology, (in Finnish)

Hjelt S-E (1973) Experiences with automatic magnetic interpretation using the thick plate model. Geophys Prosp 21: 243 - 265

Hjelt S-E (1975) Performance comparison of non-linear optimization methods applied to interpretation in magnetic prospecting. Geophysica 13: 143-166

Hochberg JE (1964) Perception. Prentice-Hall, Englewood Cliffs, NJ

Homilius J (1970) Anwendung der geoelektrischen Widerstandsmethode über Kalksteinvorkommen - Leistungsfähigkeit und Grenzen. Beih Geol Jahrb 98: 37-64

Karous M, Hjelt S-E (1983) Linear filtering of VLF dip-angle measurements. Geophys Prosp 31: 782 - 794

Moser PK (ed) (1987) A priori knowledge. Oxford Univ Press, Oxford, 222 pp

Oldenburg DW, Ellis RG (1991) Inversion of geophysical data using an approximate inversion mapping. Geophys J Int 105: 325 -353

Parasnis DS (1974) Some present-day problems and possibilities in mining geophysics. Geoexploration 12: 97 - 120

Parasnis DS (1975) Mining geophysics. Elsevier, Amsterdam, 395 pp

Parker RL (1980) The inverse problem of electromagnetic induction: existence and construction of solutions based on incomplete data. J Geophys Res 85: 4421-4428

Popper KR (1986) Objective knowledge. An evolutionary approach, revised edn. Clarendon Press, Hong Kong, 395 pp

Raiche AP (1991) A pattern recognition approach to geophysical inversion using neural nets. Geophys J Int 105: 629 - 648

Steiner F (1988) Most frequent value procedures (a short monograph). Geophys Trans Hung 34 (2-3): 139 - 260

Weidelt P (1981) Extremal models for electromagnetic induction in two-dimensional perfect conductors. J Geophys 49: 217 - 225

Weidelt P (1985) Construction of conductance bounds from magnetotelluric impedances. J Geophys 57: 191 - 206

Chapter 2

INTERPRETATION USING NOMOGRAMS

" A rule of thumb usually is true -

    at least approximately

    and in some circumstances.

It does not necessarily need to be true

    and more often than not it doesn't"

(Origo, a Finnish causerie writer, 1977)

CHAPTER 2

INTERPRETATION USING NOMOGRAMS

2.1 CHARACTERISTIC POINTS

Before the era of computers, the only way to obtain quantitative interpretation was to make use of numerical information in the form of simplified "rules of thumb" or more elegantly of nomograms. These were produced by scale model experiments or tedious calculations by hand. Since the amount of information that could be squeezed into a nomogram is limited, the information from only selected points of an anomaly profile was used. Rarely more than three or four model parameters can be determined using nomogram interpretation, which also limits the methodology to single body anomalies. Today interpretation nomograms still have their role as a backup tool in difficult situations without access to computers, as a quick, first-aid inversion in remote field conditions or as a means to produce initial parameter values for a more elaborate inversion procedure.

We consider here the construction of nomograms from measures or characteristic points of a geophysical anomaly. One can speak of horizontal measures and vertical (or amplitude) measures. In addition combinations of these can naturally be specified. Useful characteristic points or measures of an anomaly have to be selected so, that they can be easily and reliably determined from measurements. It is also an advantage, if a mathematical (theoretical) relationship exists between the characteristic points and the model parameters. In the ideal situation an error estimate ought to be available. Measures, which are affected by many parameters or by data errors, should be avoided.

When the information carried by the characteristic points or measures is concentrated into a nomogram, some basic rules are important to be followed in order to construct nomograms that work well. The curves on the nomograms should contain areas of approximately equal size and close to rectangular form, even if their "sides" can be curved. If the curves on some regions of the nomogram chart become crowded, the resolution degrades strongly. To provide for maximum amount of information in a single nomogram chart, dimensionless combinations of both the anomaly measures and the model parameters should be preferred. These items will be highlighted mainly by discussing selected examples of geophysical interpretation nomograms and to demonstrate their use.

We will introduce the systematic notations in Table 2.1 for the characteristic points or measures of an anomaly. A dash after a symbol indicates differentiation with respect to the co-ordinate. A collection of measures defined for magnetic anomaly interpretation in Fig.. 2.1 shows how inventive geophysicists can be. The collection has been redrawn from a critical review paper by Åm (1972), where after having discussed and criticized the wealth of measures the author continues by proposing some new measures! The notations defined in Table 2.1 have also been introduced in Fig.. 2.1, with the exception of the definitions of Naudy, which are good examples of combined measures.

TABLE 2.1   Measures of geophysical anomalies

<u>*HORIZONTAL MEASURES*</u>

x_M	location (horizontal co-ordinate) of the maximum amplitude F_M
x_m	location (horizontal co-ordinate) of the minimum amplitude F_m
x_k	location (horizontal co-ordinate) where the anomaly amplitude is $F_k = k \cdot F_M$
x'_M	location (horizontal co-ordinate) of the maximum slope of F_M.
x'_m	location (horizontal co-ordinate) of the minimum slope of F_M.
x_0	location (horizontal co-ordinate) of zero slope of F_M. [ie. x_M or x_m]
x'_{kM}	location (horizontal co-ordinate) where slop is $F' = k \cdot F'_M$

$d_k = x_{k1} - x_{k2}$ k-width of an anomaly = width of the anomaly where its amplitude is
$$F_k = k \cdot F_M.$$
$d_{Mm} = x_M - x_m$ horizontal distance between maximum and minimum of the anomaly
$d_0 = x_{01} - x_{02}$ distance between points where the anomaly is zero

<u>*VERTICAL MEASURES*</u>

A	total amplitude excursion = $F_M - F_m$.
F_M	the maximum amplitude
F_m	the minimum amplitude
F_k	the amplitude = $k \cdot F_M$, where $k < 1$
F'_M	maximum slope of the anomaly
F'_m	minimum slope of the anomaly
F'_{kM}	the slope of the anomaly is $F'_{kM} = k \cdot F'_M$, where $k < 1$
etc.	

<u>*COMBINED MEASURES*</u>

Any definitions are possible.

---

a.

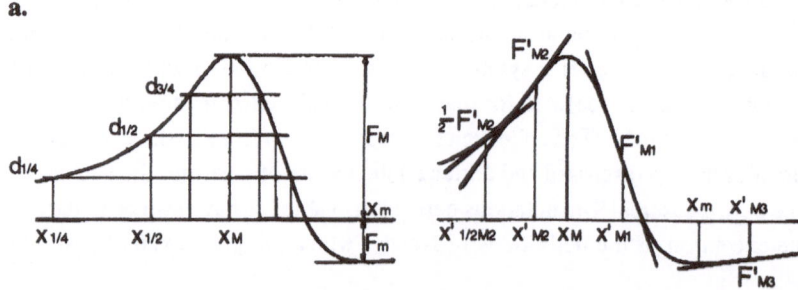

Fig. 2.1a. Definition of characteristic points and measures for magnetic interpretation using the methods of Bruckshaw and Kunaratnam (**left**) and Moo (**right**) (redrawn from Åm 1972).

b.

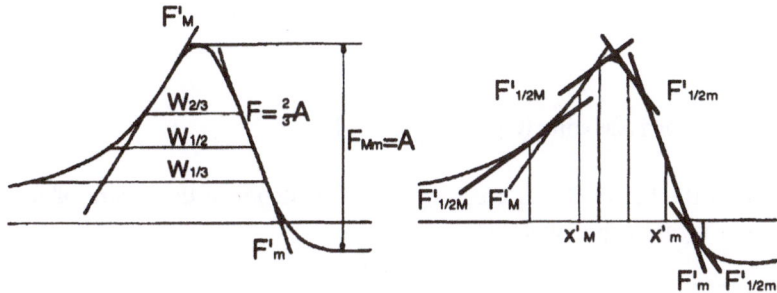

Fig. 2.1b. Definition of characteristic points and measures for magnetic interpretation using the methods of Grant and Martin (**left**) and Bean (**right**) (redrawn from Åm 1972).

c.

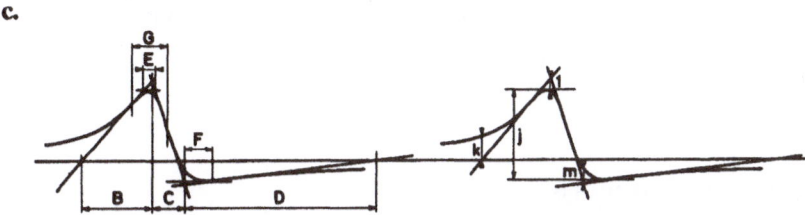

Fig. 2.1c. Definition of characteristic points and measures for magnetic interpretation using the method of Naudy (redrawn from Åm 1972). Pure horizontal and vertical measures are separated, although many of them can be considered to be of combined type.

Example of the determination and use of characteristic points

Consider the simple gravity field of a spherical body with an excess mass M. If the centre of the sphere is located at $(x = 0, z)$, then the gravity anomaly along the x-axis $(z=0)$ is

$$F = \Delta g = G\,M\frac{z}{\left[x^2 + z^2\right]^{3/2}} \tag{2.1}$$

From

$$\frac{\partial F}{\partial x} = -3\ G{\cdot}M\frac{x{\cdot}z}{\left[x^2 + z^2\right]^{5/2}} = 0 \tag{2.2}$$

the extremum values of Δg are located at $x_M = 0$ and $x_m = \infty$. The maximum amplitude is

$$F_M = \Delta g(x= 0) = G{\cdot}M/z^2. \tag{2.3}$$

Let us determine the k-width of the anomaly. First the location of x_k is solved from:

$$G \cdot M \frac{z}{\left[z^2 + x_k^2\right]^{3/2}} = k \cdot G \cdot M / z^2. \tag{2.4}$$

We have

$$x_k = \pm z \cdot \sqrt{k^{-2/3} - 1} \quad \text{and accordingly} \quad d_k = 2z \cdot \sqrt{k^{-2/3} - 1}. \tag{2.5}$$

The k-width, thus also the half-width is proportional to the depth to the centre of the sphere. Interpretation is done using the expression

$$z = d_k / [2 \cdot \sqrt{k^{-2/3} - 1}] \tag{2.6}$$

which eg. for $k = 0.5$ (the half-width) gives $z = 0.651 \cdot d_k = (\pm)1.302 \cdot x_k$. This mathematical result can be extended to other bodies as many of the "rules of thumb" of gravimetric interpretation presented in the geophysical literature over the years. By manipulating the expressions of the anomaly of sphere further equations for inversion could be developed.

2.2 NOMOGRAMS AND THEIR USE

2.2.1 Gravity Nomograms

Gravity data profiles are not very commonly interpreted using the sheet (thin plate) model in connection with the nomogram technique. It is, however, very instructive to consider two examples from a small booklet by Rao et al (1974). Th examples cover most of the typical features of nomogram interpretation. The model is defined in Fig.. 2.2 and the corresponding gravity anomaly is:

$$\Delta g = 2G \, \Delta \rho \cdot t \cdot [\sin \phi \cdot \ln(\tfrac{r_2}{r_1}) + \cos \phi \cdot (\psi_2 - \psi_1)] \tag{2.7}$$

The simplest nomogram is drawn for a vertical sheet and consists of three curves only. The model parameters are the depth to the upper and lower edge of the sheet, $p = (z_1, z_2)$. The horizontal location is exactly at the anomaly maximum, since for a horizontally symmetric body, such as this model, the gravity anomaly is also completely symmetric. The characteristic measures to be determined from the data profile are the one third- and two third-widths. One first draws a horizontal line at their ratio until it reaches corresponding curve (step 1 in Fig. 2.3). Continuing in the vertical direction from the crossing point the ratio between the model parameters is found from the horizontal axis (step 4 in Fig. 2.3). The vertical line crosses the two other nomogram curves (steps 2 and 3) and the corresponding values for the ratios $x_{1/3}{:}z_1$ and $x_{2/3}{:}z_1$ are obtained from the right vertical axis. These two values are not independent of each other, but they provide us with two values to increase the reliability of the inversion. Having now z_1, the earlier determined ratio of the depths finally gives also z_2.

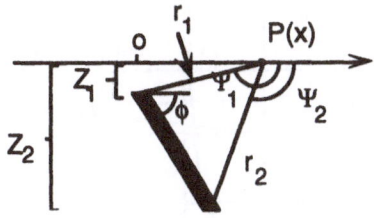

Fig. 2.2. The sheet (thin plate) model used in the example on nomogram interpretation of gravity anomalies (redrawn from Rao et al. 1974)

When inclined sheets are considered, the model parameter has one additional component, the dip; $\mathbf{p} = (z_1, z_2, i)$. One additional group of curves will be needed. From the first nomogram, using the characteristic parameters of the anomaly (steps 1 and 2, Fig. 2.4) a point is determined. The co-ordinates of this point in the nomogram curve frame (steps 3 and 4) give us the dip and the depth ratio. Entering these values into the second (additional) nomogram (Fig. 2.5) allows a value of z_1 to be determined and accordingly from the ratio obtained at step 4, finally z_2.

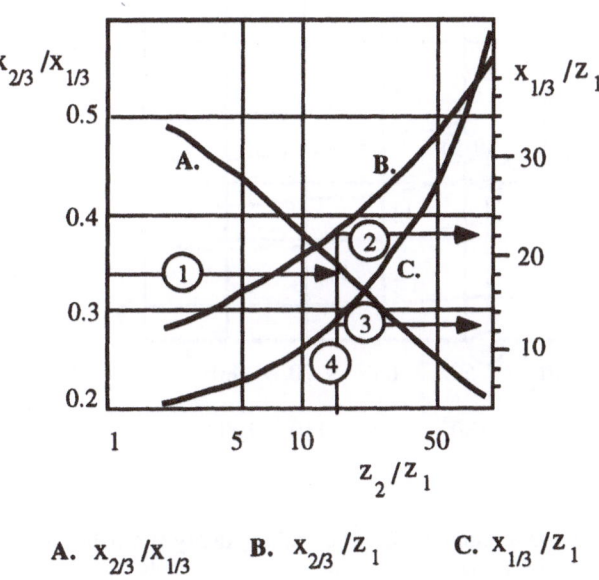

Fig. 2.3. Nomogram (called standard curves by Rao et al. 1974) for interpretation of gravity anomalies of vertical sheets.

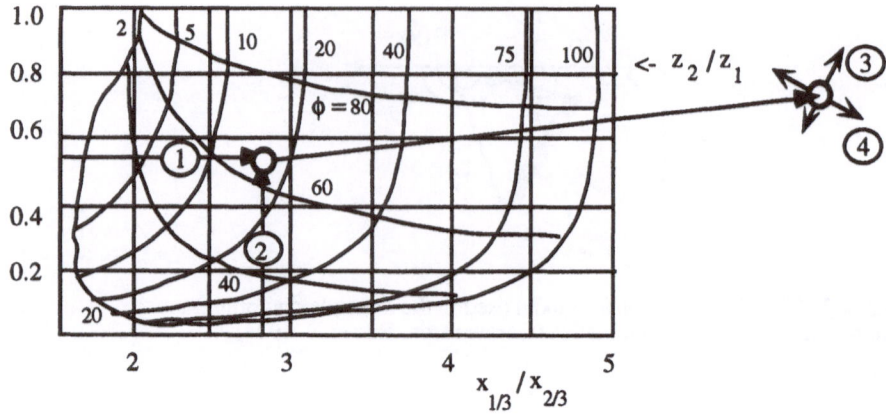

Fig. 2.4. Nomogram for first stage of interpreting gravity anomalies of inclined sheets. Steps 3 and 4 have been shifted and enlarged for the sake of clarity (after Rao et al. 1974)

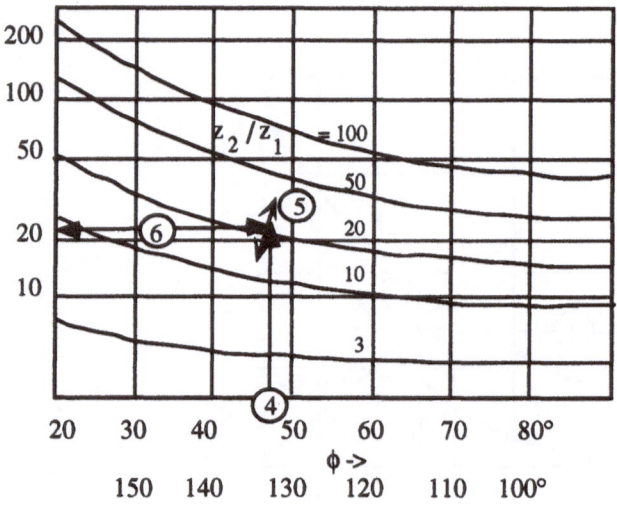

Fig. 2.5. Second stage nomogram for interpreting gravity anomalies of inclined sheets. Results from the first nomogram (Fig. 2.4.; steps 3 and 4) are introduced into this nomogram (steps 4 and 5), the last step 6 giving the upper depth and together with the result of step 4 also the lower depth (after Rao et al. 1974)

2.2.2 Nomogram for Interpretation of VLF Resistivity Data

In the VLF resistivity (or in short VLF-R) method, two horizontal components of the electric and magnetic field of the plane wave produced by a distant radio transmitter are measured. In some devices, the receiver output is directly calibrated to show the apparent resistivity (ρ_a) and the phase (ϕ), data parameters used in the magnetotelluric method. Since the trasnmitters operate in the VLF frequency band (15 - 30 kHz), the VLF-R method can be considered as a variant of controlled source magnetotellurics. Since the EM wave at a single frequency is normally received, the VLF-R method does not provide a sounding effect, although the method is capable of detecting well conducting layers in crystalline bedrock down to a depth of 200 - 300 m.

Since two data parameters (ρ_a, ϕ) are measured, in principle one is able to determine at most two model parameters at each measuring site. An often used inversion method is to assume a two-layer Earth, which can be described with three model parameters: $\mathbf{p} = (\rho_1, \rho_2, h_1)$. Keeping ρ_1 fixed or assuming its value, the two other parameters can be determined. Although the analytic expression for a plane wave two-layer EM field is not too complicated and suitable inversion schemes based on e.g. optimization could easily be developed (Hjelt et al. 1985), the Canadian producer of the VLF instrument EMR-16R has introduced a series of nomograms to assist in rapid inversion of the data in field conditions (Geonics 1972). The nomogram is easily reconstructed by using the standard magnetotelluric equations for the plane wave impedance above a two-layer Earth. A nomogram exists for a series of values of ρ_1 ranging over typical resistivities of overburden layers. An experienced

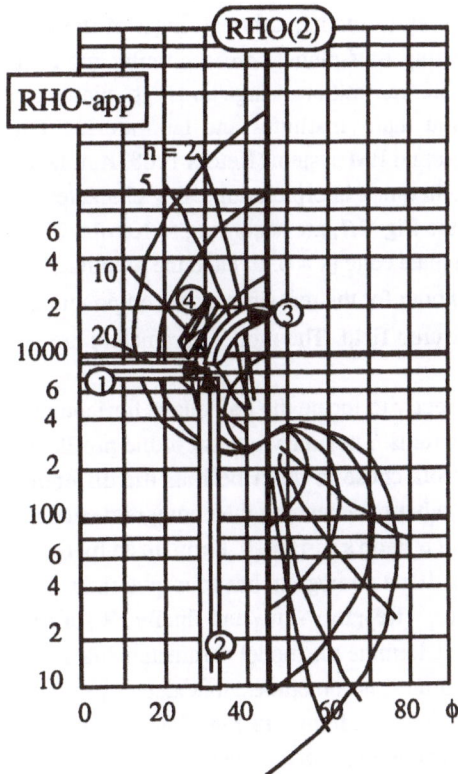

Fig. 2.6. An interpretation nomogram for the determination of two-layer Earth parameters from VLF-R data. The nomogram is for resistivity of the upper layer, 300 Ohm-m. Introducing apparent resistivity and phase from the axes defines a point on the nomogram. The co-ordinates of this point, read from the nomogram axis rho(2) and the curves for h, give the values of the interpretation parameters. (Recalculated and redrawn after Geonics 1972)

geophysicist can easily estimate from the surroundings, where the data are taken, an approximate value for ρ_1. If the data is interpreted by somebody without knowledge of ρ_1, interpretations can be made using several of the nomograms and those ρ_2-h_1-combinations are selected, for which the most systematic and reasonable variation of the parameters occur along the profile.

Figure 2.6 is a modified copy of one of these nomograms. The stages of interpretation are shown in circles numbered from 1 to 4. The axes of ρ_a and ρ_2 are in fact the same, but for the ease of use they are drawn at two separate locations. The form of the nomogram curves tells also directly about the limitations of the model at hand (i.e. overburden resistivity ρ_1). In this example depths greater than 40 m cannot be resolved, since the VLF wave is attenuated too much while penetrating through the overburden layer. Also the resolution of very thin overburden is poor. When the measured phase is 45^0 the measured apparent resistivity approaches the true resistivity of the second (bedrock) layer. In the example, the data $\mathbf{F} = \{\rho_a = 800 \ \Omega m, \phi = 32^0 \ \text{(stages 1 and 2, Fig. 2.6)}\}$ produces the interpretation (approximately) $\mathbf{p} = \{\rho_2 = 2.000 \ \Omega m, h = 14 \ m \ \text{(stages 3 and 4, Fig. 2.6)}\}$.

2.2.3 Nomograms for Magnetic Dipole (Slingram) Profiling

Modelling of two- and three-dimensional electromagnetic fields have always been a great computational problem and it is a challenge even today when supercomputers are available. Scale models have been a solution, used to overcome the computational difficulties. Model materials limit, though, the simulation of real geological situations (e.g. resistivities of the surrounding bedrock).

Geometrically complicated conductors are quite easy to model, although most of the scale models have used simple bodies. Large plates, simulating half-planes, are a popular model in electromagnetic prospecting work using the magnetic dipole transmitters. Large and high-quality sets of model profiles have been measured and published at many institutes and laboratories. One example is the set for the Slingram (or horizontal loop) ground EM system (Ketola 1968, Ketola and Puranen 1967). The set contains also nomograms for use in quick interpretation using characteristic measures of the profiles. The best measures, defined in the Fig. 2.7, are according to Ketola (1968) the location and the amplitude of the minimum above the plate edge [x = 0, F_2] and the amplitudes of the side maxima [F_1 and F_3]. These measures are defined both for the real (in-phase) component and for the imaginary (out-of-phase) component of the measured field. The model parameters are $\mathbf{p} = $ (h/a, x, S = σd and α).

Since the position of the minimum can be used directly to locate the half-plane horizontally, three parameters remain to be determined from the nomograms. The nonsymmetry of the profile has direct connection with the dip (α) of the plane and Ketola chose to describe it as the difference between the amplitudes of the side maxima (F_1 - F_3). The other parameter on the nomogram curves is the depth of the top of the half-plane below the profile on the Earth's surface, h, normalized by one of the system parameters, the transmitter-reciever distance a. Each nomogram has four quadrants, I for the combination Re_2; Im_2, II for Re_2; Re_1 - Re_3, III for Re_1 - Re_3; Im_1 - Im_3 and finally IV for Im_2; Im_1 - Im_3. Any or several of the quadrants can be used to determine the model parameters h/a and α. The resolution of the nomogram curves differ for the quadrants, is sometimes nonexistent (quadrant III in Fig. 2.8), sometimes too poor for a practical use (e.g. quadrant I in Fig. 2.8). To increase reliability, Ketola (1968) suggests using more than one quadrant to produce several model parameter estimates. If the estimates differ considerably, then the electrical parameter w, associated with that nomogram, has an incorrect value. A choice is made among the nomograms with different w, until a consistent solution has been found.

A different nomogram is produced for each electrical parameter (labelled w by Ketola), which is dimensionless

$$w = \mu\omega\sigma \, da = q \cdot S, \qquad (2.8)$$

with $q = \mu\omega$ a containing only the system parameters ω, the transmitter frequency and a, the transmitter-reciever distance. The parameter μ, the magnetic permeability, is in fact one of the electric parameters of the model, but can for most applications be considered constant ($= \mu_0$, the free space magnetic permeability). The conductance of the plate, $S = \sigma d$, is thus determined by selecting the nomogram with a proper value w as described .

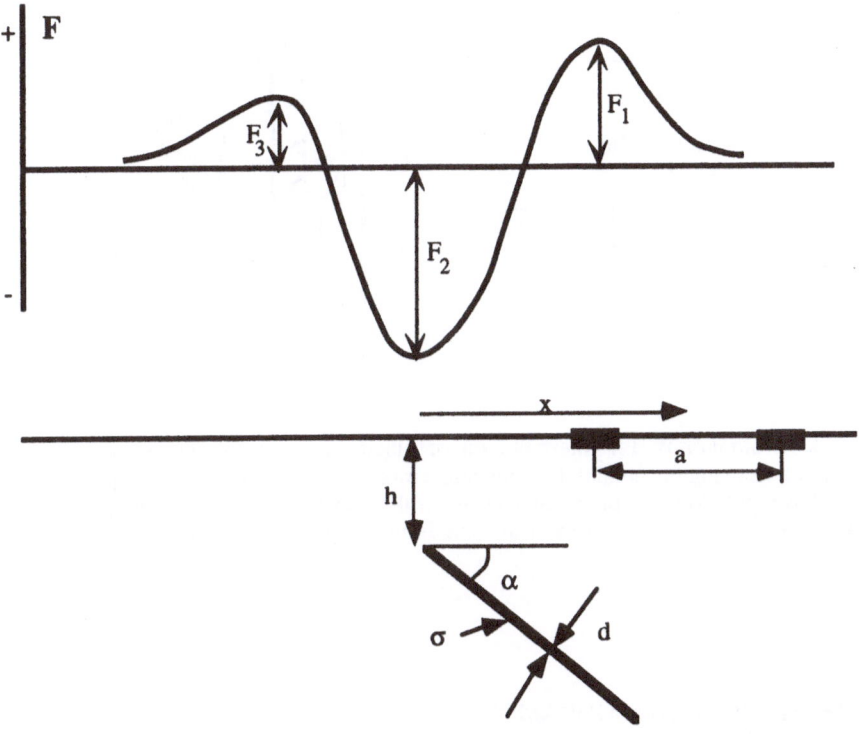

Fig. 2.7. Definition of the data parameters (**above**) and model parameters (**below**) for the half-plane model in an electromagnetic dipole (horizontal loop = Slingram) method.   F is either Re, the in-phase component or Im, the out-of-phase (quadrature) component. (from Ketola 1968)

From the measured profile one can easily deduce (the components are scaled to be in % of the amplitude of the primary field): $Re_1 = +6.9$ %; $Re_2 = -18.0$ %; $Re_3 = 0$ %; $Im_1 = 6$ %; $Im_2 = -14.6$ % and $Im_3 = 3.6$ %. From the nomogram shown in Fig. 2.8, one has the estimates $p = \{h/a = 0.15 ...0.2, x = 0, S = \sigma d = 4.5$ and $\alpha = 70... 75^0\}$. Actually $w = 5.26$ is first determined giving S; finally by assuming a thickness or 5 m from the borehole data the estimate of conductivity is 0.9 S/m. With $a = 40$ m the depth to the top of the plate is 6 ... 8 m, about the overburden thickness of the area. The dip estimate agrees well with the result obtained from two boreholes crossing the ore body (Fig. 2.9).

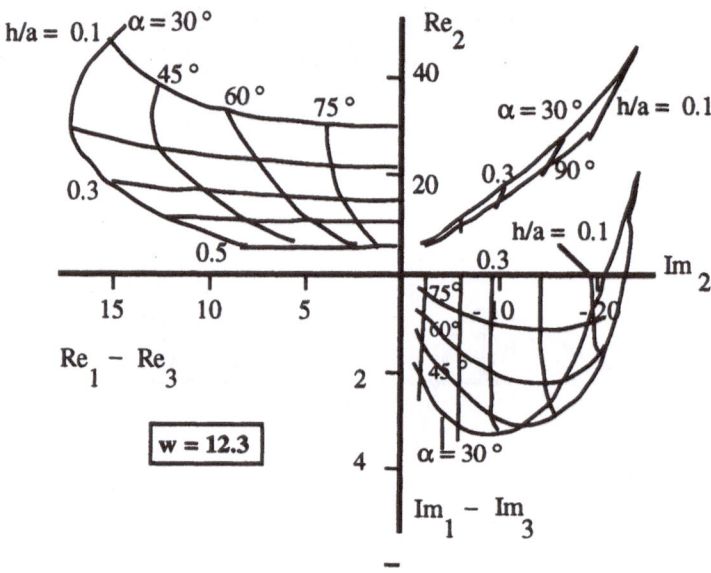

Fig. 2.8. Nomograms for interpretation of electromagnetic dipole (Slingram) measurements using the conducting half-plane model. Pairs of characteristic points are used as input. Any quadrant can be used, if the nomogram has a reasonable resolution. The data and model parameters are defined in Fig. 2.7 (from Ketola 1968)

2.3. COMPUTERIZED "NOMOGRAMS"

When extensive and high-quality model calculations or scale model data are available, the use of characteristic points and nomogram techniques are easily included in a interactive computer system. The nomograms are stored in the computer as a numerical databank (Pelton et al. 1978). A suitable numerical scheme is added to take care of interpolation both along a curve of a nomogram and across nomograms (variation of model parameters). A further, obvious step would be to store the profile data directly instead of the nomograms only.

Valkeila and Rossi (1990) have described such an application closely connected to the horizontal-loop example described. They have stored magnetic dipole (horizontal loop) scale model profiles containing 15 to 40 data points each, in a computer data bank. Their data bank consists of 940 models and about 32 000 data points covering the thin sheet (half-plane) conductors with parameters (definitions see above) h/a = 0.1 .... 0.6, α = 5°.... 175°, w = 0.9 .... 44. A two-stage interpolation consists of a parabolic fitting to three successive digitized points of each anomaly profile and a subsequent linear interpolation between the profiles. Further improvements of the system makes use of the hyperparabolic fit technique (cf. Sect. 4.3.1) and Lagrangian interpolation.

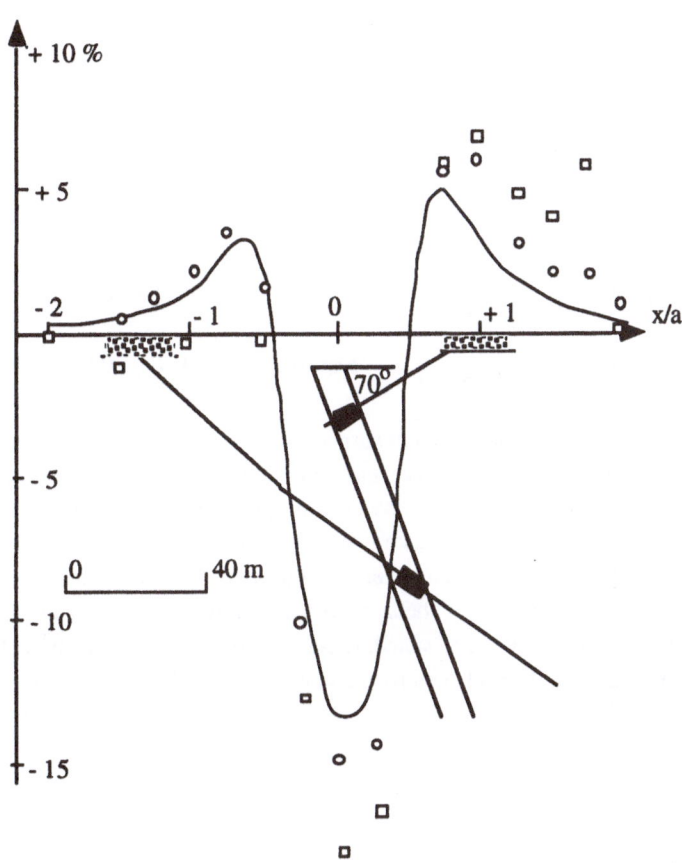

Fig. 2.9. Interpretation of horizontal loop (Slingram) data by conducting half-plane scale model curves. **Solid line:** model curve for h/a = 0.2; α = 75; w = 5.26. **Squares** measured real component data; **circles** measured imaginary component data; a = 40 m. **Black regions** borehole intersection of ore. (Redrawn from Ketola 1968)

2.4 ON THE ACCURACY OF CHARACTERISTIC POINT INVERSION

Depending on the choice of characteristic points and whether analytic expressions exist for the anomalous field produced by the model or not, the accuracy of characteristic point interpretation can be estimated to some extent. In the example on Slingram interpretation an idea about the amount of interpretation errors is obtained by comparing with existing borehole information. For models with simple analytical expressions even an elaborate error analysis is possible. Kalinina and Holzmann (1967, 1971) have presented such an analysis for the horizontal cylinder gravity measurements. They assume that Gaussian error with $e = N (0, \sigma)$; $\sigma = 0.1 \cdot F_M$ is added to the theoretical anomaly of the cylinder:

$$F_t = \frac{2mz}{x^2 + z^2} .$$
(2.9)

For three cases the results are:

a) $p = (z) = L^{-1}(F_M')$ $-> e = \Delta z = 0.55\, z$

b) $p = (m, z) = L^{-1}(F_M, d_{1/2})$ $-> e = (\Delta m, \Delta z) = (0.11\, m, 0.10\, z)$ (2.10)

and   c) $p = (m, z) = L^{-1} (F_h)|_{LSQ, N = 11}$. $-> e = (\Delta m, \Delta z) = (0.06\, m, 0.04\, z)$

The result of case a. is at first thought surprising - at least in comparison with the others. As is shown later in Chapter 8, the theoretically smallest error should be obtained when one parameter is determined alone. The poor accuracy of determining exactly the maximum slope for data with considerable error content is, however, a known fact and is indicated by this huge error: 55%. When the depth, h, and source amplitude, m are determined from the half-width and the anomaly maximum respectively, the parameter error is of the same order than the standard error of the data (10%). A slight improvement, to about half the % stated, is obtained by using the principle of least squares for a whole anomaly profile (with N = 11 data points in example c). The parameter errors are 6 and 4 % respectively.

REFERENCES

Geonics Ltd (1972) Manual for EM 16R. Geonics Techn Note TN-1.

Hjelt S-E, Kaikkonen P, Pietilä R (1985) On the interpretation of VLF resistivity measurements. Geoexploration 23: 171 - 181

Kalinina TB, Holzmann (Golt'sman) FM (1967) A statistical interpretation algorithm and its application in the solution of the inverse problem of magnetic exploration. Izv Akad Nauk SSSR Phys Solid Earth (7): 451-457

Kalinina TB, Holzmann (Golt'sman) FM (1971) Analytical comparison of methods of geophysical interpretation. Izv Akad Nauk SSSR Phys Solid Earth (5): 320-325

Ketola M (1968) The interpretation of Slingram (horizontal loop) anomalies by small-scale model measurements. Geol Survey of Finland, Rep of Investig 1, Otaniemi, 19+79 pp

Ketola M, Puranen M (1967) Type curves for the interpretation of Slingram (horizontal loop) anomalies over tabular bodies. Geol Survey of Finland, Rep of Investig 2, Otaniemi

Pelton WH, Rijo L, Swift CM Jr (1978) Inversion of two-dimensional resistivity and induced polarization. Geophysics 43: 788 - 803

Rao BSR, Radhakrishna Murty IV, Rao CV (1974) Gravity interpretation by characteristic curves. Andhra Univ Press, Waltair, 44 pp

Valkeila T, Rossi M (1990) The curves for magnetic dipoles and their use in computer interpretation. In: Hjelt S-E, Fokin AF (eds): Electrical prospecting for ore deposits in the Baltic Shield 2. Electromagnetic methods. Geol Survey of Finland, Rep of Investig 95, Otaniemi, pp 28 - 37

Åm K (1972) The arbitrarily magnetized dyke: interpretation by characteristics. Geoexploration 10: 63 - 90

Chapter 3

LINEAR PARAMETERS

" The shortest distance between two points is a line

- unless somebody has fooled you

with a non-linear scale of reference"

(Unknown practical joker)

CHAPTER 3

LINEAR PARAMETERS

3.1 DEFINING THE LINEAR PROBLEM

3.1.1 Genuine Linear Problems

A great majority of geophysical fields can be calculated from an integrals of the convolution type

$$F_t(\mathbf{r}) = \int_S [G(\mathbf{r} - \mathbf{r}_o) \cdot \mathbf{p}(\mathbf{r}_o) \, d\mathbf{r}^3]$$ (3.1)

Inversion consists of equating F with the measurements and solving for the unknown physical parameter distribution, labelled $k(\mathbf{r}_o)$ in Fig. 3.1. Since the model parameter appears linearly in the equation, then defining the boundary S will retain the linearity of the parameter p.

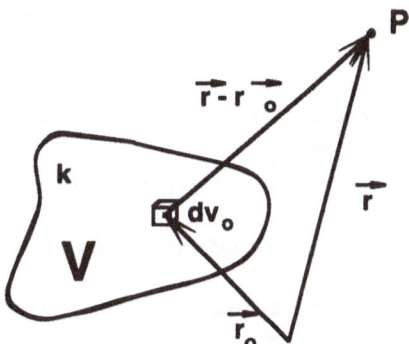

Fig. 3.1. Calculation of the potential field of a material parameter k. The field point is P, k varies within the volume ("model body") V.

Example 1. Potential fields

The gravity field of a body with *density* ρ $(\mathbf{r}_o)$ is:

$$g_t(\mathbf{r}) = \nabla \int_V [\frac{1}{(\mathbf{r} - \mathbf{r}_o)} \cdot \rho(\mathbf{r}_o) \, d\mathbf{r}^3] = \int_S [\frac{1}{(\mathbf{r} - \mathbf{r}_o)^3} \cdot \rho(\mathbf{r}_o) \, d\mathbf{r}^2].$$ (3.2)

The magnetic field of a body with *magnetization* $M(\mathbf{r}_o)$ is correspondingly:

$$H_t(r) = \nabla \int_V \{\nabla \cdot [\frac{1}{(r-r_o)} \kappa(r_o)] \, dr^3\} \qquad (3.3)$$

Example 2. Electromagnetic sounding, EMS (e.g. Hohman and Raiche 1988)

The electric field is calculated as a sum of two terms, the E_n being the field which would exists for a homogeneous(or layered) Earth.

$$E_t(r, f) = E_n(r, f) + \int_S [G(r_o, r, f) \cdot E(r_o, f) \, \sigma_a(r_o, f) \, d S_o] \qquad (3.4)$$

A similar experssion can be written for the magnetic field, with different Green's function, **G**, naturally. The kernel of the integral, the Green's function, contains also information on the type of source field used in a specific EMS method (plane wave, line source = long cable, magnetic dipole source = "small" loop).

3.1.2 Discretization of the Linear Model

If the *parameter* (p as in the general definition, k as in the 1D example to follow, ρ as the density, κ as the magnetic susceptibility or σ_a as the electrical conductivity in the examples above) is constant within S, one can take *param*(r_o) outside the integral and hope to evaluate $\int_S [G(r-r_o) \, dr_o^3]$ in closed form. The commonly used alternative is to model the Earth by piecewise constant regions S_j (cf. Figs. 3.2 and 3.3 for the one-dimensional parameter distribution, where the symbol k is used for the physical parameter and the symbols S and S_j are replaced by the lengths L and L_j). The calculation of F_t then becomes:

$$F_{ti}(r_i) = \sum_j k_j(r_o) \int_{S_j} \{G_j(r_i - r_{oj}) \, dr_j^3\} = \sum_j k_j G_{ij} \qquad (3.5)$$

In vector notations and returning to use the general symbol p instead of the physical parameter k:

$$F_t(r) = G(r - r_o) \, p(r_o). \qquad (3.6)$$

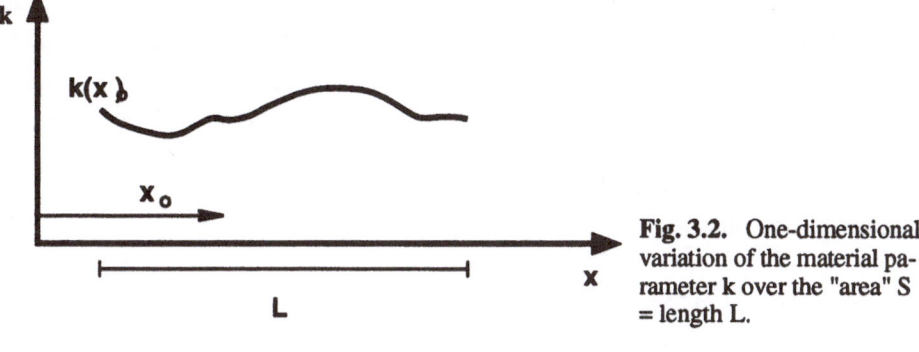

Fig. 3.2. One-dimensional variation of the material parameter k over the "area" S = length L.

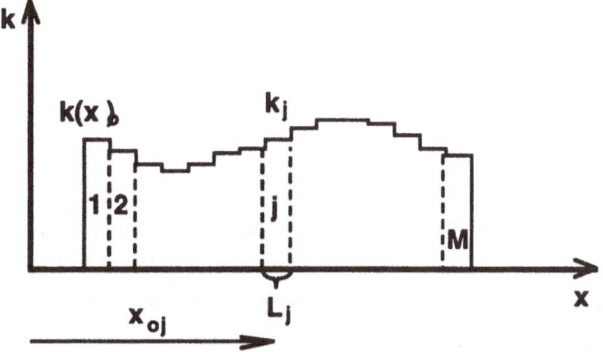

Fig. 3.3. The digitized version of the material parameter distribution k(x), which is assumed to be constant over each interval.

3.1.3 Linearization

A great part of the geophysical model parameters appear in a non-linear form. There are three main alternatives to linearize the expression for a geophysical field and its inversion:

> A. Combination of variables;
>
> B. Transformation of variables;
>
> C. Taylor series approximations.

In alternative A, a combination of variables can be identified, which linearize the problem. Let us take as an example magnetic inversion using a special technique, which has been named Werner-deconvolution by Hartmann et al. (1971). The magnetic anomaly of a thin, vertically infinite magnetized sheet at field point (x_i, z_i) is:

$$F_i = F(x_i, z_i) = (A x + B z)/[x^2 + z^2] \tag{3.7}$$

$$x = x_i - x_o, \text{ and } z = z_i - z_o,$$

with $p = (A, B, x_0, z_0)$, A and B being the components of magnetization along and perpendicular to the dip of the sheet and (x_0, z_0) the co-ordinates of the upper edge of the sheet. The parameters A and B are linear, x and z non-linear ones. Assume for simplicity (but without loss of generality) that the measuring profile is on the surface $z = 0$. From Eq. (3.7) one finds by multiplication:

$$F_i [x_i^2 - 2x_ix_0 + x_0^2 + z_0^2] = A x_i - A x_0 + B z_0 \qquad (3.8a)$$

By rearranging one obtains:

$$F_i x_i^2 = a_0 + a_1 x_i - b_0F_i + b_1F_i x_i , \qquad (3.8b)$$

with:

$a_0 = B z_0 - A x_0$

$a_1 = A$

$b_0 = - [x_0^2 + z_0^2]$

$b_1 = 2x_0$

Since x_i are known F_i for all values of i, the four intermediate parameters a_0, a_1, b_0 and b_i can be determined from a linear equation system with four equations using data F_i at four field points x_i.

The solved linear intermediate parameters can be transformed into the original model parameters by inverting equations to give $p = L\{F_i\} = L_{interm}\{a_0, a_1, b_0, b_1\}$ in the form:

$x_0 = b_1/2$

$z_0 = [-b_0 - (b_1/2)^2]^{1/2}$

$A = a_1 \qquad (3.9)$

$B = (a_0 + a_1b_1/2) / [-b_0 - (b_1/2)^2]^{1/2}$

This method, orginally proposed by Werner (1955), was redefined by Hartmann et al. (1971), who applied the operator in Eq. (3.9) successfully to sequences of four data points along aeromagnetic profiles.

This approach lead to useful inversion algorithms (interpretation operators) only when there exists algebraic inverse relations between the intermediate variables and the original (non-linear variables).

The best known (and perhaps mostly used) example of the linearization alternative B, the transformation of variables, is resistivity sounding. The classical way of approaching DC sounding curves has been to draw them on log-log paper. The interpretation proceeds by shifting curves up and down and sideways along the coordinate system. The success of this technique relates to the linearity of the logarithm of the variables. Figure 3.4 shows "how to straighten the Banana Valley" from an early paper by Johansen (1977) on computerized DCS inversion.

As a third alternative of linearization function series approximations can be used. Any function can be expanded (given certain conditions) into a Taylor series:

$$F(p, x_i) = F(p_0, x_i) + \Sigma \partial F/\partial p_j (p_0, x_i) \Delta p_j + \Sigma\Sigma \partial^2 F/\partial p_j\partial p_k (p_0, x_i) \Delta p_j \Delta p_k + .. \quad (3.10)$$

Neglecting terms of higher order the vector of the parameter changes is linear $\Delta p = p - p_0$. This forms the basis of many iterative inversion methods in geophysics.

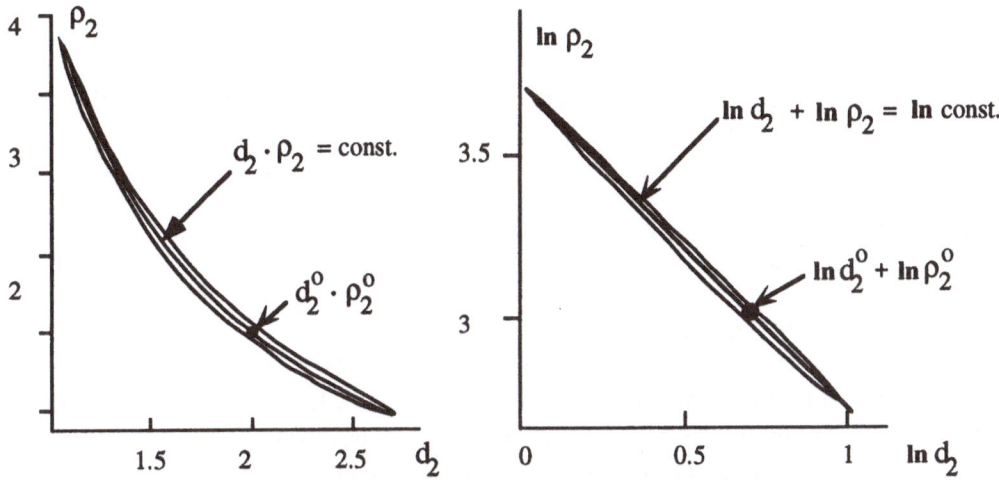

Fig. 3.4. The linearization of model parameters by transformation of variables. To the left the original parameters, the (**curved**) line of the product of the parameters = const. and the banana-shaped contour the 68% confidence limit of the parameters, when the data error is 5 %. To the **right** the same for logarithmically transformed (= linearized) parameters. The product **curve** has become a **line**. The parameters are the thickness (horizontally) and resistivity (vertically) of a three layer model in DC Schlumberger sounding. (After Johansen 1977)

3.1.4 Iterative Approaches

The use of series expansions to linearize the calculation of the fields and find an inverted model is explained through an example from electromagnetic inversion. The linearization uses Fréchet derivatives, an exact and fashionable notation replacing the traditional concept of partial derivatives. A functional equation relates small changes in conductivity $\Delta\sigma$ to small changes in data ΔF_h

$$\Delta F_{hi} = \int \{D_i[\sigma(z), z] \, \Delta\sigma(z)\} \, dz \qquad i = 1, 2, ...N_{data} = N_h. \quad (3.11)$$

Terms of second order are neglected and the D_i are called Fréchet kernels. Since the kernels are functions of conductivity, the inverse problem for σ is still nonlinear but linear for the changes $\Delta\sigma$. Every iterative solution of $\Delta\sigma$ is a linear inverse problem.

The Fréchet kernels for HEMD (= horizontal electromagnetic dipole loop) methods are (Hohmann and Raiche 1988):

$$D_E(r, z, f) = \int_0^\infty [\frac{E^2(\lambda, z, f)}{Ia} J_1(\lambda a)] \, \lambda \, J_1(\lambda r) \, d\lambda$$

$$D_r(r, z, f) = Q^* \int_0^\infty D_E(r, z, f) \, \lambda^2 \, J_1(\lambda r) \, d\lambda \tag{3.12}$$

$$D_z(r, z, f) = -Q^* \int_0^\infty D_E(r, z, f) \, \lambda^2 \, J_1(\lambda r) \, d\lambda$$

with $\qquad Q = I/(i \, \omega \, \mu_0)$

and where I is the current in the source loop, a the loop radius, D_E the Fréchet kernel for the electric field, D_r the kernel for the horizontal component of the magnetic field and D_z the kernel for the vertical component of the magnetic field. If a piecewise continuous conductivity (the layered Earth model) is assumed, then the equation for the unknown conductivities $\Delta \sigma_j$ of N_m layers:

$$\Delta d_i = \Sigma_j D_{ij} \Delta \sigma_j \qquad\qquad i = 1,2, ...N_h; \, j = 1, 2, ... \, N_m., \tag{3.13a}$$

has to be solved iteratively. The discrete kernels D_{ij} are

$$D_{ij} = \int_{\Delta z_j} D_i(\sigma, z, dz) \, . \tag{3.13b}$$

It has been found useful to normalize both sides of the basic Eq. (3.13a) by the standard deviation of the error of each datum. Such a procedure de-emphasizes observations with large uncertainties.
Each step of the iteration consists of:

1. Calculation of the theoretical response $F_t^{(k)}$ for the conductivities $\sigma_j^{(k)}$ $j = 1, 2, ... N_m$ to produce $\Delta d_i = F_{hi} - F_{ti}^{(k)}$
2. This is substituted into the Eq. (3.13a) which is solved for the $\Delta \sigma_j^{(k+1)}$.
3. $\sigma^{(k+1)} = \sigma_j^{(k)} + \Delta \sigma^{(k+1)}$

A good initial guess for the conductivities $\sigma_j^{(0)}$ $j = 1, 2, ... N_m$ will be necessary to start the iteration.

3.2 FITTING LINEAR MODELS TO DATA

3.2.1 Systems of Linear Equations

The geophysical linear inversion problem is mathematically the solution of a system of linear equations. Depending on whether *exact fit* (the number of data points and model parameters are equal) or a *least-squares fit* (the number of data points is significantly greater than the number of model parameters) is attempted, the system of equations is *determined* or *overdetermined*. In practice also the *underdetermined* case (the number of data points is less than the number of model parameters) is often encountered. An overdetermined or determined system can turn into an underdetermined one, if the matrix is singular. This can easily happen because of data errors.

The solution of the linear equation

$$\mathbf{F}_h(r) = \underline{\mathbf{G}}\,\mathbf{p}, \tag{3.14}$$

is sought in the form

$$\mathbf{p} = \underline{\mathbf{H}}\,\mathbf{F}_h(r). \tag{3.15}$$

The quality of the solution can be monitored by studying the resolution matrix:

$$\underline{\mathbf{R}} = \underline{\mathbf{H}}\,\underline{\mathbf{G}}, \tag{3.16}$$

which is the mapping from the space of all possible solution vectors $\mathbf{p}^*$ into one selected solution $\mathbf{p}$. R equals to the identity matrix:

$$\mathbf{I} = \begin{bmatrix} \begin{matrix} 1\ 0 \\ 0\ 1 \end{matrix} & 0 & 0 \\ 0 & \begin{matrix} \cdots\ 0 \\ 0\ \cdots \end{matrix} & 0 \\ 0 & 0 & \begin{matrix} 1\ 0 \\ 0\ 1 \end{matrix} \end{bmatrix}$$

for an unique solution.

Another way of studying the solution is by using the information (density) matrix

$$\underline{\mathbf{S}} = \underline{\mathbf{G}}\,\underline{\mathbf{H}}, \tag{3.17}$$

which describes the independence of the components of the data vector $\mathbf{F}_h$. One has:

$$\mathbf{F}_{model} = \underline{\mathbf{S}}\,\mathbf{F}_h, \tag{3.18}$$

and we have a complete fit of the model field to the data, if $\underline{S} = \underline{I}$. The interpreter usually strives at a minimum parameter variance . For independent data, the variance can be written as:

$$\text{var}(p) = E(p - E(p))^2 = \sum_{j}^{N} H_{ji}^2 \, \text{var}(F_{hj}).$$

(3.19)

The properties and size of the various linear systems of equation is demonstrated graphically in Fig. 3.5, where F_h is the vector of data, M the number of model parameters, N the number of data points, p the vector of model parameters and $\underline{G}$ the coefficient matrix, the structure of which will depend both on the choice of model and the method of fit. The term k stands for the rank of $\underline{G}$ (= the greatest dimension of the non-singular partial matrix of $\underline{G}$). The existence of solutions of the linear problem is demonstrated graphically (following an idea of P. Weidelt) for the case of two unknowns in Fig. 3.6.

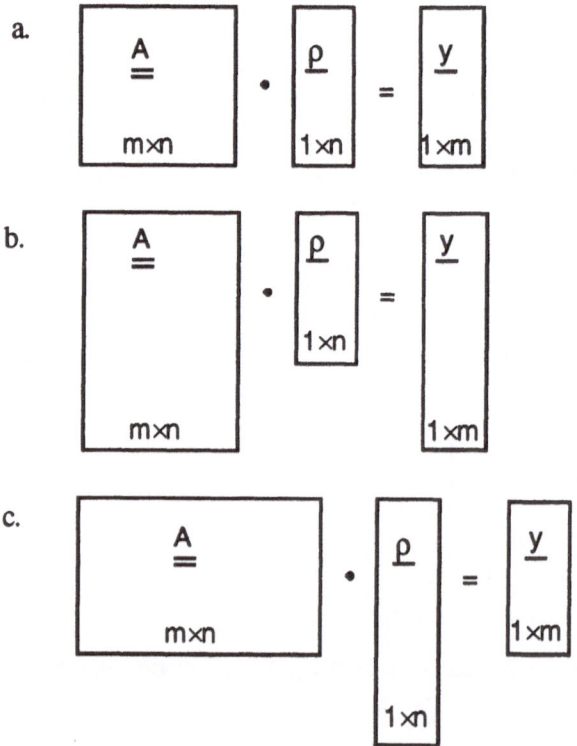

Fig. 3.5. Three types of systems of linear equations. **a.** Determined: # of unknowns = # of equations. **b.** Overdetermined: # of unknowns < # of equations. **c.** Underdetermined: # of unknowns > # of equations.

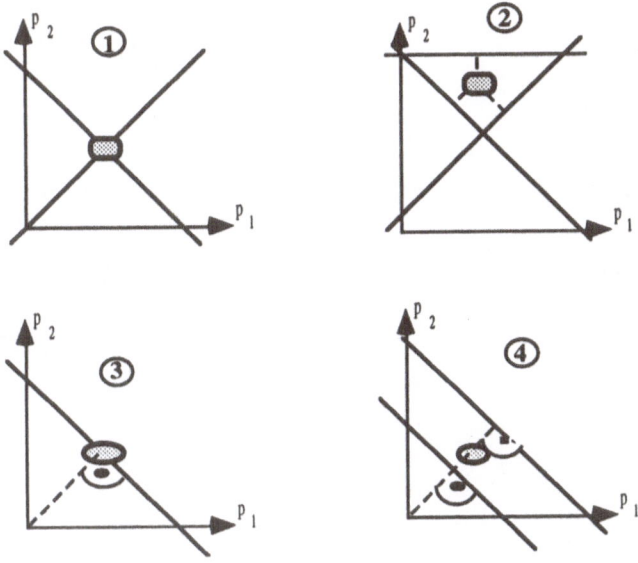

Fig. 3.6. Graphical description of linear equations with two unknowns. (After Weidelt 1978).

The four equations are:

1. $M = N = k = 2$ $\begin{aligned} p_1 + p_2 &= 1 \\ p_1 - p_2 &= 0 \end{aligned}$ Determined system   A unique solution exists

2. $M = k = 2; N = 3$ $\begin{aligned} p_1 + p_2 &= 1 \\ p_1 - p_2 &= 0 \\ p_3 &= 1 \end{aligned}$ Overdetermined system   Only an LSQ solution exists

3. $M = 2; N = k = 1$ $p_1 + p_2 = 1$ Underdetermined system ∞ number of solutions [a]

4. $M = N = 2; k = 1$ $\begin{aligned} p_1 + p_2 &= 1 \\ p_1 + p_2 &= 0.5 \end{aligned}$ $\begin{aligned} p_1 + p_2 \text{ is overdetermined} \\ p_1 - p_2 \text{ is underdetermined} \end{aligned}$ No solution

[a] the general solution is $x = 0.5 \, (1, 1)^T + \alpha \, (1, -1)^T$, where α can be chosen arbitrarily.

The division of linear systems into determined, over- and underdetermined has been described in the textbook of Lanczos (1961) and for geophysical applications by Bott (1973).

3.2.2 Exact Fit

In this case one writes:

$$F_h(r) = F_t(r,p) = \underline{G}\,p$$

$$\qquad\qquad (3.20)$$

$$F_{ti}(r_i) = \Sigma\,G_{ij}\,p_j = G_i\,p\,.$$

The solution is simply:

$$p = [\underline{G}]^{-1}\,F_h\,. \qquad\qquad (3.21)$$

3.2.3 Least Squares (LSQ)

Now:

$$S = min = \Sigma e^2 = \Sigma[F_h(r) - F_t(r)]^2 = e\cdot e^T. \qquad\qquad (3.22)$$

If $F_t(r)$ depends linearly on the model parameters, one of the advantages of the LSQ method is that it retains this linearity. This is easily shown. For LSQ:

$$min\,S \rightarrow \quad \partial S/\partial p = 0 = \Sigma[F_h(r) - F_t(r)]\{-2h\cdot\partial F_t/\partial p\}. \qquad\qquad (3.23)$$

Alternatively in vector notations:

$$0 = (F_h - \underline{G}\,p)\,\underline{G}^T,$$

from which:

$$p = [\underline{G}\,\underline{G}^T]^{-1}\,\underline{G}^T\,F_h\,. \qquad\qquad (3.24)$$

Additional details concerning exact fit, least squares solutions and robust estimation techniques [which deal with the problem of strongly erroneous data ("outliers")] will be described in section 4.1.1.

3.3 ITERATIVE SOLUTIONS: THE METHOD OF GAUSS-SEIDEL

Although solution of the linear (or linearized) equations can be obtained by a great variety of methods, it may be instructive to study in more detail some of the most efficient ones, the Gauss-Seidel iterative method. In section 3.4 another powerful approach to solving the basic linear equations will be discussed.

The linear system of equations (3.14):

$$\mathbf{F_h} = \underline{G}\, \mathbf{p}\,,$$

or in component form:

$$F_{hi} = \Sigma\, G_{ij}\, p_j$$

is to be solved for the unknown linear parameters p_j.

The principle of the Gauss-Seidel iterative method is, that from each of the equations, one of the unknowns, p_j, is solved as a function of all the others. When this is subsequently repeated for all the equations and unknowns, one has completed one iteration of the method. Let $p_j^{(k)}$ be the value of the jth unknown at the kth iteration:

$$p_j^{(k+1)} = [F_{hi} - \sum_{j=1,\ \neq i}^{N} G_{ij} p_j^{(k)}]\, /\, G_{ii} \tag{3.25a}$$

This is the original Gauss method, only after a complete iteration all the new values of the unknowns are substituted and a new iteration starts. The improvement of Seidel was always to use the latest available upgrade of the unknowns:

$$p_j^{(k+1)} = [F_{hi} - \sum_{j=1}^{i-1} G_{ij} p_j^{(k+1)} - \sum_{j=i+1}^{N} G_{ij} p_j^{(k)}]\, /\, G_{ii} \tag{3.25b}$$

This will speed up the convergence of the iterative solution. Further improvements in speed of iteration are obtained by using the multiplication factor, the optimal value of which is problem dependent.

Because the Gauss-Seidel method is an iterative one some starting values for the parameters are needed. It can be shown, that the method converges towards the correct solution from any starting point, if the matrix $\underline{G}$ is diagonally dominant, i.e.

$$|G_{ii}| \geq \sum_{j \neq i} |G_{ij}|$$

Experiments in connection with magnetic and gravity interpretation using the so called block model (Hjelt 1976) it was found that additional improvement in the speed of convergence is obtained if an appropriate starting value selection is made. Three possibilities were tested:

$$p_i^{(0)} = 0$$
$$p_i^{(0)} = F_{hi} / G_{ii}$$
$$p_i^{(0)} = F_{hi} / NG_{ii}$$

The second alternative means that, the information of the measuring point i is attributed completely to the corresponding model parameter; the last one that the information of the ith data point is divided equally among all the parameters. The optimal selection obviously is somewhere between the second and the third choice. The third choice seemed to give the best convergence for the greatest number of cases tested.

The use of the Gauss or the Gauss-Seidel (G-S) method is especially appropriate for nume-rical solution of most geophysical fields, since the numerical equivalents of wave equations and Laplace/Poisson equations are diagonally dominant.

Hjelt (1976) has used the diagonal dominance to calculate the minimum distance between elements of a block model for the G-S method to be convergent from **arbitrary** initial starting values of the linear parameters. This leads to rather great distances. Smaller distances certainly are possible, but would require extra care in selecting initial values for the iteration, since the convergence would not be guaranteed. The analysis does give some qualitative insight into the resolution properties of the G-S method, but it **does not** give the absolute minimum between model bodies.

A simple example of linear parameter modelling is given in Fig. 3.7 and Table 3.1. The (constructed) model consists of three thin, infinitely deep and inclined magnetized plates used to interpret three equi-spaced measurements of the vertical magnetic field. The magnetic susceptibi-lities k_i of the plates are the linear model parameters to be found. The geometric part of the anomaly is of the basic anomaly type I. In case (a) the edges of the plates are located at the data points, in case (b) the edges are shifted so, that the maximum anomaly of each plate is situated at the data points. In case (c) the data points are located closer to each other than the plates.

Starting from 0 susceptibility of all plates, the solution converges towards the correct values (parameter k of plate 1 only, shown in Fig. 3.7) in cases (a) and (b), although neither has a strictly diagonally dominant matrix. In case (c) no convergent solution is found, not even by restricting the susceptibility to positive values, as is physically appropriate. Case (b) is closer to diagonal domi-nance and it has clearly the best convergence properties.

Table 3.1 also contains some other information, which is often used to describe the inversion properties of a linear matrix. The column following the matrix elements gives the sensitivity of the susceptibilities to changes in the measured field, ΔF. Evidently small sensitivity values are associated with a good stability of the inversion. A low condition number is stated to be a require- ment for stable inversion of a matrix. This clearly singles out case (c) of this example as the worst one, but the condition number does not seem to rank the convergence properties of cases (a) and (b) properly. That the inversion of a matrix is unstable, if the determinant of the matrix is small, seems to hold true for this example. The quotient between the smallest and largest eigenvalues of the matrix is also indicative of how well a matrix can be inverted (see Sect. 3.4).

TABLE 3.1 Properties of the coefficient matrix $\underline{G}$ of a three-parameter inversion example. (Hjelt, 1976)

	$\underline{G}$			$\Delta p/\Delta F$	λ_G	$\lambda_{min}/\lambda_{max}$	cond ($\underline{G}$)	Det ($\underline{G}$)
(a)	1.0	1.0	0.6	2.61	$1.257 + j\,0.57$			
	0.0	1.0	1.0	2.52	$1.257 - j\,0.57$	0.352	2.84	0.92
	0.2	0.0	1.0	1.52	0.486			
(b)	1.209	0.805	0.586	1.74	1.756			
	0.308	1.209	0.805	1.30	1.355	0.295	3.39	1.23
	0.167	0.308	1.209	1.42	0.518			
(c)	0.4	1.0	0.483	21.7	$0.582 + j\,0.713$			
	-0.154	0.4	1.0	10.0	$0.582 - j\,0.713$	0.041	24.9	0.0317
	-0.207	-0.154	0.4	4.0	0.0374			

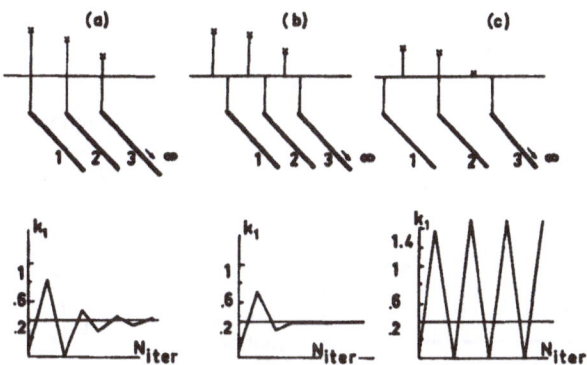

Fig. 3.7. **Top row** Three thin, infinitely deep, magnetized plates and the location of three data points (vertical magnetic field) used for linear inversion of the magnetic susceptibilities k_i of the plates. **Bottom** Convergence of the susceptibility k_1 of the first plate for Gauss-Seidel iteration. The three variations (a), (b) and (c) are explained in the text. (Hjelt 1976).

3.4 GENERALIZED INVERSION

We consider again the linear parameter problem

$$F_h = \underline{G}\,p \tag{3.14}$$

where $\mathbf{p}$ is the vector (length M) of model parameters, $\mathbf{F_h}$ the data vector (length N) and $\underline{\mathbf{G}}$ the (N x M) model matrix as discussed. The equation can be considered as a linear mapping from the parameter space to the data space (Fig. 3.8). The inverse problem, the determination of the parameter values from the data or construction of the interpretation operator, consists of finding the matrix $\underline{\mathbf{H}}$, which performs an inverse mapping from the data space to the parameter space (Fig. 3.8)

$$\mathbf{p} = \underline{\mathbf{H}}\,\mathbf{F_h} \tag{3.15}$$

Classically, the equation has a solution only for M=N. If the model matrix $\underline{\mathbf{G}}$ is well behaved and non-singular, there are no problems in determining the matrix $\underline{\mathbf{H}}$. For many real world problems, however, this often is not the case, even for M=N. The generalized inverse approach using the singular value decomposition technique has become an important tool in handling linear (or linearized) geophysical inverse problems. The important development of the generalized inverse approach is that also cases M≠N can be handled.

For this purpose the matrix $\underline{\mathbf{G}}$ is written as the product of three matrices, one of which is the diagonal matrix $\underline{\Lambda}$ consisting of the eigenvalues of the matrix $\underline{\mathbf{G}}$ and the two others consist of the corresponding eigenvectors. We will, for simplicity, consider first the case where M=N and after that extend to the more general case. A great variety of other definitions of $\underline{\mathbf{H}}$ than the SVD presentation to follow can be found in Bjerhammer (1973).

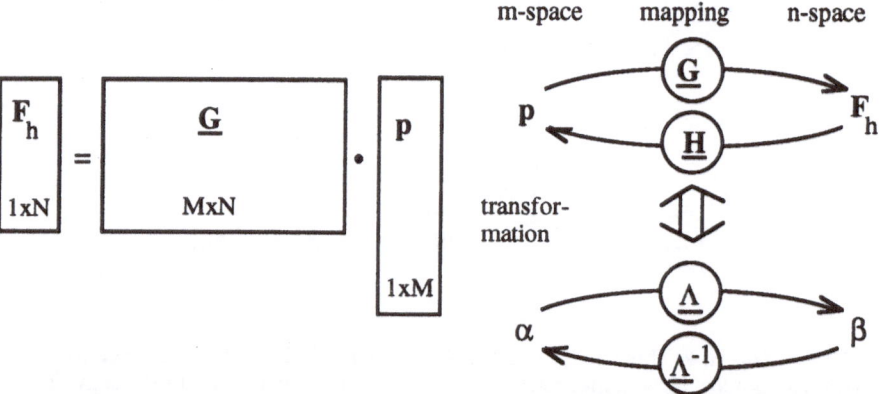

Fig. 3.8. In a system of linear equations the parameter space is mapped into the data space and vice versa. Transformation of parameters and data into their generalized counterparts α and β retains the mapping properties.

3.4.1 Singular Value Decomposition (SVD)

Consider the equation of determining the eigenvalues λ of $\underline{\mathbf{G}}$ of an MxM square matrix:

$$\underline{\mathbf{G}}\,\mathbf{u} = \lambda\,\mathbf{u} \tag{3.26}$$

where **u** is the eigenvector corresponding to the eigenvalue λ. Since there (in the non-degenerate case) are M eigenvalues and correspondingly M eigenvectors, they all combine into:

$$\underline{G}\,\underline{U} = \underline{U}\,\underline{\Delta}\ . \tag{3.27}$$

It is easy to show that if two eigenvalues $\lambda_i \neq \lambda_j$, then corresponding eigenvectors $\mathbf{u}_i$ and $\mathbf{u}_j$ are perpendicular to each other. If the eigenvectors are normalized (their length $\| \mathbf{u}_i^2 \| = 1$), then:

$$\underline{\tilde{U}}\,\underline{U} = \underline{U}\,\underline{\tilde{U}} = \underline{I},$$

where $\underline{I}$ is the identity matrix.

By multiplying the matrix eigenvalue equation from the right with $\underline{\tilde{U}}$ one obtains:

$$\underline{G}\,\underline{U}\,\underline{\tilde{U}} = \underline{G} = \underline{U}\,\underline{\Delta}\,\underline{\tilde{U}},$$

which gives the required expression of $\underline{G}$ as a product of three matrices. Furthermore:

$$F_h = \underline{G}\ p = \underline{U}\,\underline{\Delta}\,\underline{\tilde{U}}\,p\ . \tag{3.28}$$

Multiplying from the left with $\underline{\tilde{U}}$

$$\beta = \underline{\tilde{U}}\,F_h = \underline{\tilde{U}}\,\underline{U}\,\underline{\Delta}\,\underline{\tilde{U}}\,p = \underline{\Delta}\,\underline{\tilde{U}}\,p = \underline{\Delta}\,\alpha, \tag{3.29}$$

where two new vectors,

$$\alpha = \underline{\tilde{U}}\,p \qquad \text{and} \tag{3.30}$$

$$\beta = \underline{\tilde{U}}\,F_h,$$

are introduced. They are called the generalized parameter and data vectors respectively (Wiggins 1972, in one early presentation on the subject talks about reparametrization). One sees immediately that:

$$\alpha = \underline{\Delta}^{-1}\,\beta \tag{3.31}$$

which is an extremely simple case of linear mapping between two spaces (cf. Fig. 3.8). In the non-singular case this solution is easy to construct, since the eigenvalue matrix $\underline{\Delta}$ is a diagonal matrix and its inverse just has the inverse values of the corresponding eigenvalues on its diagonal. Furthermore, using the definitions of α and β:

$$\underline{\tilde{U}}\,p = \underline{\Delta}^{-1}\,\underline{\tilde{U}}\,F_h,$$

and by multiplying the result form the left with $\underline{U}$, we have finally found the inverse matrix $\underline{H}$:

$$\underline{U}\,\tilde{\underline{U}}\,p = p = \underline{U}\,\Delta^{-1}\,\tilde{\underline{U}}\,F_h = \underline{H}\,F_h .$$

Summarizing:

$$\underline{G} = \underline{U}\,\Delta\,\tilde{\underline{U}} ,$$ (3.32a)

and

$$\underline{H} = \underline{U}\,\Delta^{-1}\tilde{\underline{U}},$$ (3.32b)

showing the beautiful symmetry of the singular value decomposition for a square matrix.

3.4.1.1 Zero Eigenvalues

Suppose that p first of the eigenvalues of the matrix $\underline{G}$ differ from zero and the rest N-p of the eigenvalues are zero. Then:

$$\Delta = \begin{bmatrix} \lambda_1 & & & & & \\ & \lambda_2 & & & \underline{O} & \\ & & \ddots & & & \\ & & & \lambda_p & & \\ & & & & 0 & \\ & \underline{O} & & & & 0 \end{bmatrix} = \begin{bmatrix} \Delta_p & \underline{O} \\ \underline{O} & \underline{O} \end{bmatrix}$$ (3.33)

with $\Delta_p = \begin{bmatrix} \lambda_1 & & & \underline{O} \\ & \lambda_2 & & \\ & & \ddots & \\ \underline{O} & & & \lambda_p \end{bmatrix}$ being a matrix of size p x p.

Then it follows, that:

$$\underline{U}\,\Delta = \begin{bmatrix} \underline{U}_p\,\Delta_p & \underline{O}_{N-p} \end{bmatrix}$$

$$\underline{G} = \underline{U}\,\Delta\,\tilde{\underline{U}} = \begin{bmatrix} \underline{G}_p & \underline{O} \\ \underline{O} & \underline{O} \end{bmatrix},$$

and

$$\underline{G}_p = \underline{U}_p\,\Delta_p\,\tilde{\underline{U}}_p .$$ (3.34)

Splitting up the generalized data vector β into two parts, a vector of length p and another of length

N-p:

$$\beta = \begin{bmatrix} \beta_p \\ \beta_{N-p} \end{bmatrix} = \begin{bmatrix} \Delta_p & \underline{0} \\ \underline{0} & \underline{0} \end{bmatrix} \alpha .$$

(3.35)

This equation can be solved in the general case only, if $\beta_{N-p} \equiv 0$. This means that:

$$\beta_k = \left(\tilde{U}_k F_h \right)_k = 0 \qquad\qquad k = p+1, p+2, ..., N ,$$

which is true only, if $U_k = u_k \perp F_h$. This imposes restraints on the data set, i.e. the data vector has to be perpendicular to the eigenvectors which belong to the zero eigenvalues. This is an unrealistic requirement. In addition, we have:

$$G\, u_k = \lambda_k\, u_k = 0 \qquad\qquad \text{for } k = p+1, p+2, ..., N.$$

Thus, in the case of zero eigenvalues only p generalized parameters can be solved for; the rest N-p can be chosen freely. This fact is directly true also for the original parameters.

3.4.1.2 Zero Eigenvalues, a Least-Squares Approach

It was shown earlier that for a least-squares solution of a linear parameter problem, one has to solve the equation:

$$\tilde{G}\, G\, p = \tilde{G}\, F_h .$$

(3.36)

Substituting the singular value decomposition for G and multiplying the resulting equation from the left with $\tilde{U}$, the equation for the generalized parameter and data becomes:

$$\Delta\, \tilde{U}\, F_h = \Delta^2\, \tilde{U}\, p \qquad\qquad \text{or} \quad \Delta\, \beta = \Delta^2\, \alpha .$$

(3.37)

Since now the data vector F_h is multiplied with Δ, where automatically $\Delta_{N-p} \equiv 0$, the equation is satisfied without additional orthogonality constraints for the data.

The generalized (Lanczos-) inverse is defined as:

$$\Delta^{-1} = \begin{bmatrix} \Delta_p^{-1} & \underline{0} \\ \underline{0} & \underline{0} \end{bmatrix} ,$$

(3.38)

for which evidently holds:

$$\Delta\, \Delta^{-1} = \Delta^{-1}\, \Delta = \begin{bmatrix} I_p & \underline{0} \\ \underline{0} & \underline{0} \end{bmatrix} .$$

(3.39)

By writing:

$$\alpha = \alpha_p + \alpha_o \qquad\qquad \text{and} \quad \beta = \beta_p + \beta_o, \qquad\qquad (3.40)$$

with
$$\alpha_p^T = [\alpha_1, \alpha_2,..., \alpha_p, O_{N-p}^T],$$
$$\alpha_o^T = [O_p^T, \alpha_{p+1}, \alpha_{p+2}..., \alpha_N],$$
$$\beta_p^T = [\beta_1, \beta_2,....., \beta_p, O_{N-p}^T],$$
and
$$\beta_o^T = [O_p^T, \beta_{p+1}, \beta_{p+2}..., \beta_N],$$

and similarly:

$$\underline{U} = \underline{U}_p + \underline{U}_o,$$

one has the following identities:

$$\Delta \, \alpha_o = \Delta \, \beta_o = \underline{O}$$

$$\underline{U}_p \, \alpha_o = \underline{U}_o \, \alpha_p = \underline{O}$$

$$\widetilde{\underline{U}_o} \, \underline{U}_p = \widetilde{\underline{U}_p} \, \underline{U}_o = O$$

$$\widetilde{\underline{U}_p} \, \underline{U}_p = \begin{bmatrix} I_p & \underline{O} \\ \underline{O} & \underline{O} \end{bmatrix}$$

$$\widetilde{\underline{U}_o} \, \underline{U}_o = \begin{bmatrix} \underline{O} & \underline{O} \\ \underline{O} & I_{N-p} \end{bmatrix}.$$

The solution to the linear inverse problem then becomes:

$$p = \underline{U}_p \, \alpha_p = \underline{U}_p \, \Delta^{-1} \, \beta_p . \qquad\qquad (3.41)$$

Since α_o can be chosen freely, the obvious choice is $\alpha_o = O$, leading to the identity:

$$\underline{U}_p \, \alpha_p \equiv \underline{U} \, \alpha .$$

Since furthermore $\Delta^{-1}\beta = \Delta^{-1}\beta_p + \Delta^{-1}\beta_o = \Delta^{-1}\beta_p$, the final result is:

$$p = H \, F_h \qquad\qquad \text{with} \quad H = \underline{U}\Delta^{-1}\widetilde{\underline{U}} = \underline{U}_p \, \Delta_p^{-1}\widetilde{\underline{U}_p} . \qquad\qquad (3.42)$$

The choice $\alpha_o = O$ has a direct connection to the length of the parameter vector $\|p\|^2$. Expanding and making use of the identities above, one gets:

$$\|\underline{p}\|^2 = \|\underline{U}_p\alpha_p + \underline{U}_o\alpha_o\|^2 = (\widetilde{\alpha_p}\widetilde{\underline{U}_p} + \widetilde{\alpha_o}\widetilde{\underline{U}_o})(\underline{U}_p\alpha_p + \underline{U}_o\alpha_o) = \widetilde{\alpha_p}\alpha_p + \widetilde{\alpha_o}\alpha_o, \qquad (3.43)$$

which shows that the selected choice minimizes the length of the parameter vector. This is a useful and important property for iterative solutions, where in fact not the parameters themselves, but their linear changes Δp are solved. The minimal length of p helps to keep the solution always within the range of linearity. Using the same expansions, the minimum value of the objective function can also be calculated

$$S = \|\epsilon\|^2 = |F_h - \underline{G}\ p|^2 = \|\underline{U}_p\beta_p + \underline{U}_o\beta_o - \underline{U}_p\Delta_p\alpha_p\|^2 = \tilde{\beta}_o\beta_o. \qquad (3.44)$$

The result reveals also that the objective function will approach 0 only in the rare case that $\beta_o = \tilde{\underline{U}}_oF_h = 0$. The appearance of $\lambda = 0$ means physically that the linear parameter combination associated with this eigenvalue is poorly (not at all) represented in the data of the original system of equations.

3.4.1.3 Why Are Small Eigenvalues Dangerous?

Consider the generalized parameter and data vectors α and β:

$$\alpha = \tilde{\underline{U}}\ p$$

$$\beta = \tilde{\underline{U}}\ F_h$$

It is easy to estimate the variance of the generalized data vector. Since the vectors forming the matrix $\underline{U}$ are normalized $[\|u_k\| = 1]$, every component must be smaller than or equal to 1. Thus:

$$\text{var}(\beta_k) = \sum_{i=1}^{N} u_{ik}^2\ \text{var}(y_i) < N\ \max_i\{\text{var}(y_i)\} \qquad (3.45)$$

$$\text{var}(\beta_k) = \lambda_k^2\ \text{var}(\alpha_k)$$

whence
$$\rightarrow \text{var}(\alpha_k) = \text{var}(\beta_k)\ /\ \lambda_k^2$$

If one or several of the eigenvalues are small, it effectively determines the variance of the generalized parameter vector.

Similarly, as for the generalized data vector, the variance of the true parameters can be estimated. Since also the rows of the eigenvector matrix are normalized, the parameter variance becomes proportional to the inverse of the square of the eigenvalues.

$$\text{var}(p_k) = \sum_{i=1}^{N} u_{ki}^2\text{var}(\alpha_k) \approx 1/\lambda_k^2. \qquad (3.46)$$

Thus, if one or several of the eigenvalues of the matrix $\underline{G}$ are small, the corresponding parameter variances will grow disproportionally. A simple numerical example: for $\Delta F_k = 0.5\ \%$; $\lambda_k = 0.01$ -> $\Delta p_k = 5000\ \%$.

The situation of extreme parameter variances can be avoided by using the formalism of generalized inversion we presented. If one of several eigenvalues are very small (in comparison with the other ones), one distinctly puts these eigenvalues to zero. This sacrifices some of the parameters. A measure of this sacrifice is the resolution matrix, which now is degenerated and differs from the ideal identity matrix.

The resolution matrix is:

$$\mathbf{R} = \mathbf{H}\,\mathbf{G} = \mathbf{U}_p\widetilde{\mathbf{U}_p} \neq \mathbf{I}\,. \tag{3.47}$$

Similar changes are noted in the information density matrix

$$\mathbf{S} = \mathbf{G}\,\mathbf{H} = \mathbf{U}_p\widetilde{\mathbf{U}_p} \neq \mathbf{I}. \tag{3.48}$$

Consider the following example:

$$\mathbf{G} = \begin{bmatrix} 1 & 0 & 0 & 0 \\ 0 & 1 & 0 & 0 \\ 0 & 0 & 1/2 + e & 1/2 - e \\ 0 & 0 & 1/2 - e & 1/2 + e \end{bmatrix} \qquad \text{with } e = 0.0005$$

$$\mathbf{F}_h = [\,1\,2\,3\,4\,]^T$$

$$\text{var}\,(Fi) = 0.01 \qquad\qquad \text{for } i = 1...4$$

The eigenvalues are obtained as a solution of the fourth degree equation:

$$|\mathbf{G} - \lambda \mathbf{I}| = 0 = (1-\lambda)^2\,[(1/2 + e - \lambda)^2 - (1/2-e)^2].$$

The result is $\lambda = (1, 1, 1, 2e = 0.001)$. The eigenvector equation $\mathbf{G}\,\mathbf{u} = \lambda\,\mathbf{u}$ is solved for $\mathbf{u}$ by substituting each of the λ's in turn. From $\lambda = 2e$, one can easily deduce that $u_1 = u_2 = 0$, $u_3 = -u_4$. Since the length of the vector has to be 1, finally $2u_3^2 = 1$. This gives $\mathbf{u_4}^T = (0,0, 1/\sqrt{2}, - 1/\sqrt{2})$.

For the three first eigenvalues (all equalling 1), the eigenvectors have to fulfill either of the condi-tions $2u_i^2 = 1$ or $4u_i^2 = 1$; $i = 1,2,3,4$. Combining the results into a matrix gives:

$$\underline{U} = \begin{bmatrix} +1/2 & +1/\sqrt{2} & +1/2 & 0 \\ -1/2 & +1/\sqrt{2} & -1/2 & 0 \\ +1/2 & 0 & -1/2 & +1/\sqrt{2} \\ +1/2 & 0 & -1/2 & -1/\sqrt{2} \end{bmatrix}$$

$$\underline{\beta} = \tilde{\underline{U}} \, F_h = \begin{bmatrix} 3 \\ 3/\sqrt{2} \\ -4 \\ -1/\sqrt{2} \end{bmatrix} \qquad \underline{\alpha} = \Delta^{-1} \beta = \begin{bmatrix} 3 \\ 3/\sqrt{2} \\ -4 \\ -1000/\sqrt{2} \end{bmatrix}$$

and

$$p = \underline{U}\,\alpha = \begin{bmatrix} 1 \\ 2 \\ -993/2 \\ +1007/2 \end{bmatrix} \qquad \mathrm{var}\,(p) = \begin{bmatrix} 0.01 \\ 0.01 \\ \approx 5000 \\ \approx 5000 \end{bmatrix} \qquad \underline{R} = \underline{Iu}\,.$$

The effect of the small eigenvalue is seen as the instability of the solution p_3 and p_4 and the huge variances of both these parameters. The resolution matrix itself is an identity matrix in this case.

If one wishes to cope with the variance problem, this can be done by making $\lambda = 0$. The generalized inversion then gives:

$$p = \begin{bmatrix} 1 \\ 2 \\ +7/2 \end{bmatrix} \qquad \mathrm{var}\,(p) = \begin{bmatrix} 0.01 \\ 0.01 \\ 0.01 \end{bmatrix} \qquad \underline{R} = \begin{bmatrix} 1 & 0 & 0 & 0 \\ 0 & 1 & 0 & 0 \\ 0 & 0 & 1/2 & 1/2 \end{bmatrix}$$

The resolution matrix indicates that the average of the parameters p_3 and p_4 can in fact be obtained from the data with a good variance.

One can also consider a least-squares solution of three parameters to the four data points. In this case the result gives a sum of the parameters p_3 and p_4.

$$p = \begin{bmatrix} 1 \\ 2 \\ 7 \end{bmatrix} \qquad \mathrm{var}\,(p) = \begin{bmatrix} 0.01 \\ 0.01 \\ 0.02 \end{bmatrix}$$

$$\underline{R} = \begin{bmatrix} 1 & 0 & 0 \\ 0 & 1 & 0 \\ 0 & 0 & 1 \end{bmatrix} \qquad \underline{S} = \begin{bmatrix} 1 & 0 & 0 & 0 \\ 0 & 1 & 0 & 0 \\ 0 & 0 & 1/2 & 1/2 \\ 0 & 0 & 1/2 & 1/2 \end{bmatrix}$$

For the model then:

$$F_{model} = \underline{S}\,F_h = [1 \;\; 2 \;\; 3.5 \;\; 3.5]^T.$$

3.4.2 SVD of a Rectangular Matrix

Extending the SVD analysis and generalized inversion further let us now assume that the number of parameters, M, differs from that of the number of data points, N. To determine the eigenvalues, we define another equation, which is an adjunct to the original equation:

$$\mathbf{F_h} = \underline{\mathbf{G}} \, \mathbf{p}$$
$$\mathbf{q} = \underline{\widetilde{\mathbf{G}}} \, \mathbf{x},$$

(3.49)

where $\mathbf{q}$ and $\mathbf{x}$ are arbitrary vectors of lengths M and N respectively. If both equations are joined together, we have:

$$\underline{\mathbf{S}} \, \mathbf{y} = \mathbf{z}$$

(3.50)

or
$$\left[\begin{array}{cc} \underline{\mathbf{0}}_{NxN} & \underline{\mathbf{G}}_{NxM} \\ \underline{\widetilde{\mathbf{G}}}_{MxN} & \underline{\mathbf{0}}_{MxM} \end{array} \right] \left[\begin{array}{c} \mathbf{x}_N \\ \mathbf{p}_M \end{array} \right] = \left[\begin{array}{c} \mathbf{F}_N \\ \mathbf{q}_M \end{array} \right],$$

with the subscripts indicating the sizes of each individual partial matrix or vector. (The subscript h used for the data vector earlier is omitted for the clarity). $\underline{\mathbf{S}}$ is square and self-adjoint matrix [i.e. $\underline{\mathbf{S}} = \underline{\widetilde{\mathbf{S}}}$] of the size (M+N) x (M+N). The eigenvalue equation is:

$$\underline{\mathbf{S}} \, \mathbf{w} = \lambda \, \mathbf{w} \qquad \text{with } \mathbf{w}_{M+N} = \left[\begin{array}{c} \mathbf{u}_N \\ \mathbf{v}_M \end{array} \right].$$

(3.51)

In terms of the original matrix $\underline{\mathbf{G}}$, we have two eigenvalue equations:

$$\underline{\mathbf{G}} \, \mathbf{v} = \lambda \, \mathbf{u}$$
$$\underline{\widetilde{\mathbf{G}}} \, \mathbf{u} = \lambda \, \mathbf{v}$$

(3.52)

which are not independent, since the two vectors $\mathbf{u}$ and $\mathbf{v}$ appear in both equations. The eigenvalues are said to be "shifted". Since the complete eigenvectors $\mathbf{w}$ are orthogonal to each other, it follows that:

$$\widetilde{\mathbf{w}}_i \, \mathbf{w}_k = \widetilde{\mathbf{u}}_i \, \mathbf{u}_k \pm \widetilde{\mathbf{v}}_i \, \mathbf{v}_k = 0 \qquad\qquad i \neq k.$$

This can be true for arbitrary indices and vectors only if vectors $\mathbf{u}$ and $\mathbf{v}$ form orthogonal sets separately:

$$\widetilde{\mathbf{u}}_i \, \mathbf{u}_k = 0$$

$$i \neq k$$

$$\widetilde{\mathbf{v}}_i \, \mathbf{v}_k = 0 \, .$$

It is possible to eliminate the coupling in the eigenvalue equation by multiplying the first one from the left by $\underline{\widetilde{\mathbf{G}}}$ and the second one by $\underline{\mathbf{G}}$. The result is two eigenvalue equations, one for $\mathbf{u}$ and another for $\mathbf{v}$:

$$\underline{G}\,\underline{\widetilde{G}}\,u = \lambda^2\,u$$

(3.53a)

$$\underline{\widetilde{G}}\,\underline{G}\,v = \lambda^2\,v\,.$$

The eigenvectors **u** have the length N and are called the **data eigenvectors** and the eigenvalues of the first equation the data eigenvalues. The solutions to the second equation give the parameter eigenvalues and the corresponding **parameter eigenvectors** v have the length M. The coupled eigenvalue equations extend to become:

$$\underline{G}\,\underline{V} = \underline{U}\,\underline{\Delta}$$

(3.53b)

$$\underline{\widetilde{G}}\,\underline{U} = \underline{V}\,\underline{\widetilde{\Delta}}$$

Multiplying the first of the equations from the right with $\underline{\widetilde{V}}$, the singular value decomposition of $\underline{G}$ is found to be:

$$\underline{G} = \underline{U}\,\underline{\Delta}\,\underline{\widetilde{V}}$$

(3.54)

For a symmetric matrix $\underline{G}$ the eigenmatrices $\underline{U}$ and $\underline{V}$ are equal. Proceeding as in the case of the square matrix, the situation with small or zero eigenvalues can be handled and the generalized inverse defined. We need only to take care of the distinction between the two eigenmatrices $\underline{U}$ and $\underline{V}$. The results are summarized as:

$$\underline{G} = \underline{U}\,\underline{\Delta}\,\underline{\widetilde{V}}$$

(3.54a)

$$\underline{H} = \underline{V}_p\,\underline{\Delta}_p^{-1}\,\underline{\widetilde{U}}_p$$

(3.54b)

$$\alpha = \underline{\widetilde{V}}\,p$$

(3.54c)

$$\beta = \underline{\widetilde{U}}\,F_h$$

(3.54d)

$$R = \underline{H}\,\underline{G} = \underline{V}_p\,\underline{\widetilde{V}}_p$$

(3.54e)

$$\text{var}\,(p_k) = \sum_{i=1}^{N}\{v_{ki}^2\,[\sum_{j=1}^{M} u_{ji}^2\,\text{var}\,(y_j)]\,1/\lambda_i^2\}$$

(3.54f)

3.4.3 Geophysical Examples of the Use of SVD

3.4.3.1 Gravity (Pedersen, 1977)

In the first example we consider the determination of the interface between the crustal bedrock and the upper mantle (= the Moho boundary) from gravity measurements (Pedersen 1977). The principal model, which Pedersen (1977) applies to SVD interpretation of both gravity and magne-

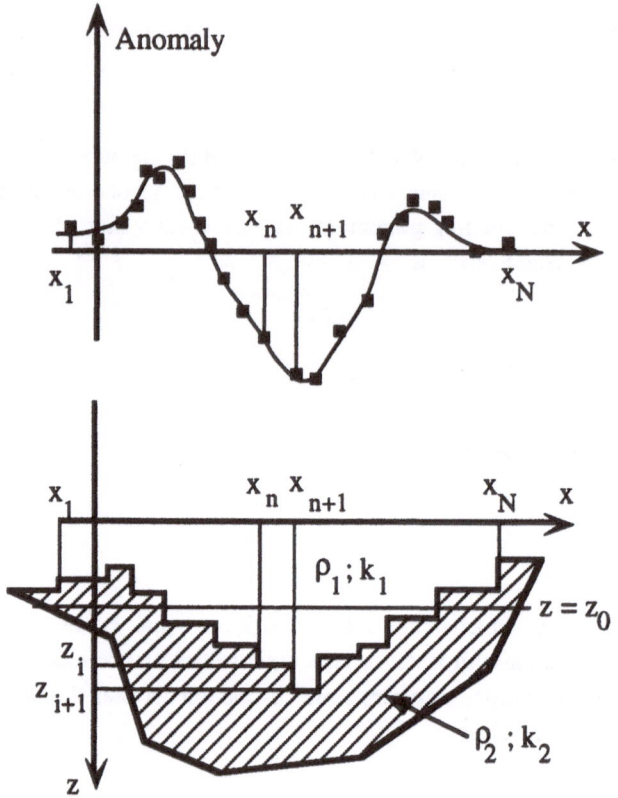

Fig. 3.9. Model for stepwise change of bedrock/upper mantle topography. **Top**: anomalous profile; **bottom**: model. (Pedersen 1977)

a.

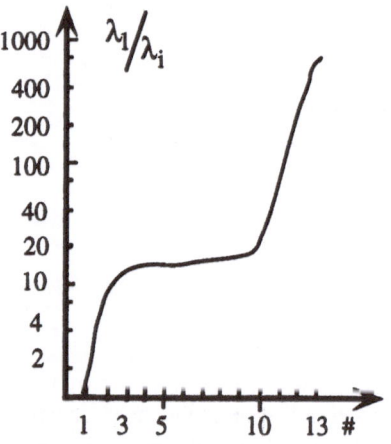

(Fig. 3.10 continues)

(Fig. 3.10 continued)

b.

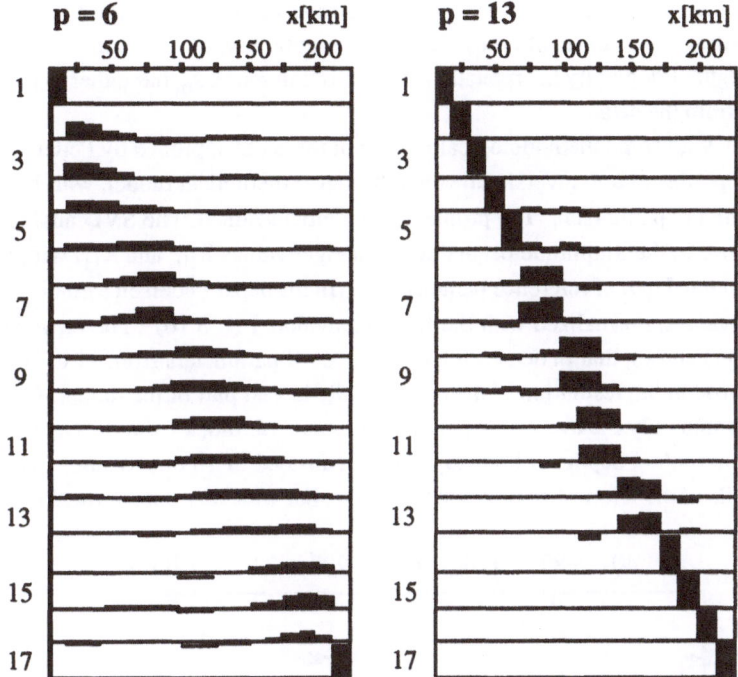

c.

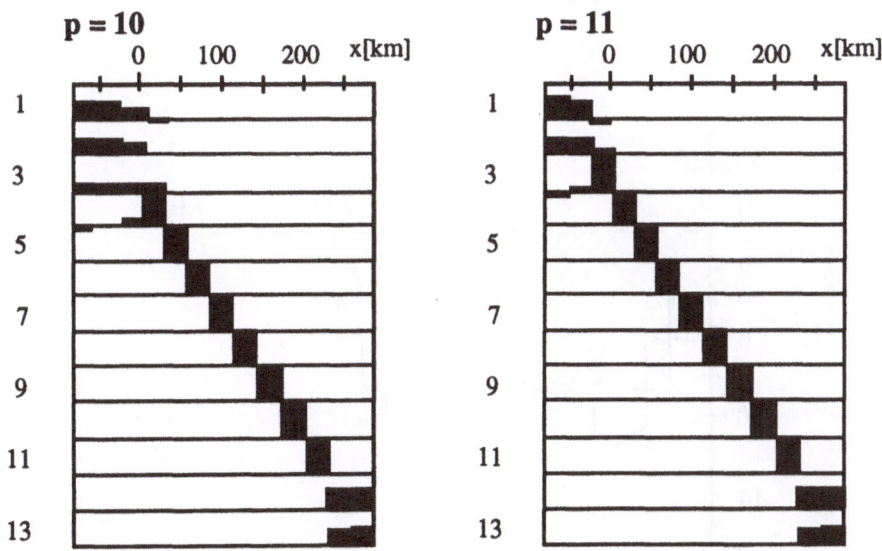

Fig. 3.10. a. (on the previous page) Inverse, normalized eigenvalues of the gravity inversion model in Fig. 3.9. **b.** Resolution matrices. **c.** Information density matrices. The number of eigenvalues, p, is given at each matrix. (from Pedersen 1977)

tic data, is shown in Fig. 3.9. The measured anomaly is discretized at points, with even spacings not necessary. The field calculated from the model is shown as squares. The model consists of stepwise changing topography of the interface between two materials, the steps occurring at equispaced distances Δx_i. The model parameters are the vertical coordinates of the stepwise changes (z_i; see Fig. 3.9). The density contrast $\Delta \rho$ or the contrast of magnetic susceptibility, Δk, is assumed constant over the whole profile length. For gravity interpretation, a depth of reference, z_o, has either to be specified or to be determined from the data.

Let us first stay with some methodological aspects of the SVD approach by Pedersen (1977). He discusses the properties of the inversion method through a theoretical model, with $N = 13$ data points and $M = 17$ model parameters. The problem is underdetermined. The SVD analysis shows that a sudden decrease in the amplitude occurs between eigenvalues λ_{10} and λ_{11}, with no significant influence from the depth of reference parameter z_o. (In his paper Pedersen actually shows the inverse of the eigenvalues, normalized with the first eigenvalue, Fig. 3.10). The error-resolution pattern (also Fig. 3.11) shows, that in decreasing the number of parameters from 17 to 10 (from M to p) a sacrifice is made in the resolution of the details in the lowest part of the model (from $x = 20$ to 170 km). For example only the average depth of the 4th and 5th block from the right can be resolved, but not the individual depths of both. By using 11 instead of 10 eigenvalues (parameters) some additional details of the leftmost slope begin to emerge, but with very poor vertical resolution.

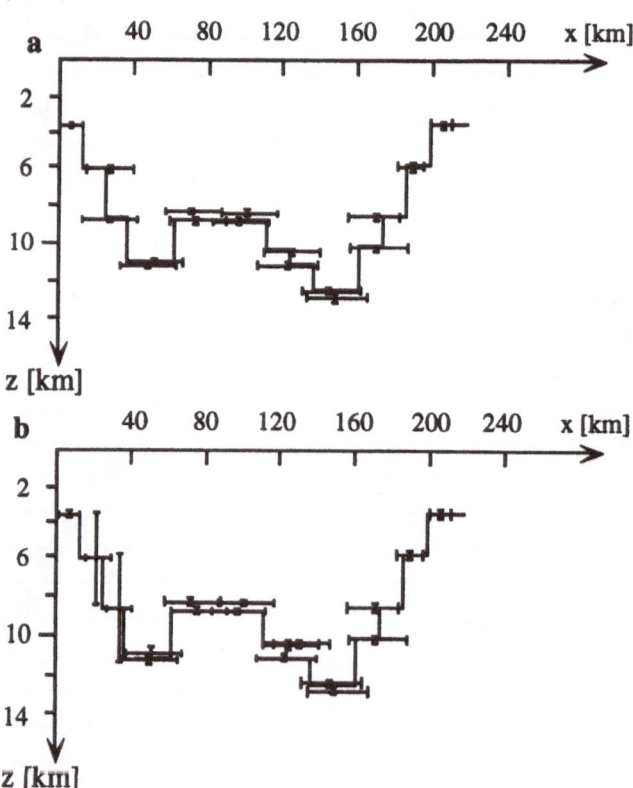

Fig. 3.11. Inverted models with error resolution bars for theoretical topography model. **a.** Full range of eigenvalues (M = 17), **b.** Reduced number of eigenvalues (p = 10). (Pedersen 1977)

The behaviour of the resolution for various number of parameters in the SVD is nicely demonstrated by a graphic exposition of the resolution matrix **R** (Fig. 3.10b.). In the ideal matrix all diagonal elements should be 1 and the off-diagonal elements 0. In Fig. 3.10, where 1 corresponds to the distances between the horizontal lines, the matrices depart remarkably from the ideal case. The smaller the number of parameters, the less is the detailed information on the topography of the bottom parts of the structure (bedrock boundary). The information density matrices **S** for the 10 and 11 parameter cases (Fig. 3.10c) shows that the third data point from the left (x = 0) is crucial in bringing in new information to the model.

To demonstrate the use of the SVD method on real data, Pedersen chose a 600 km long profile of pendulum measurements from the Mt. Desert (Worzel 1965), which was first interpreted by Oldenburg (1974). The bedrock was overlaid by sediments and water. The Bouguer profile in Fig. 3.12 is the result for removing the effects of the seawater and the sediments. The accuracy of this correction procedure, dominated by mistakes in sediment thickness, has been postulated by Pedersen (1977) to introduce an rms (root-mean-square) error of 5 mgal, distributed equally over the whole profile. Following Oldenburg, a density contrast of 430 kg/m^3 between the crystalline rocks of the crust and the mantle was used.

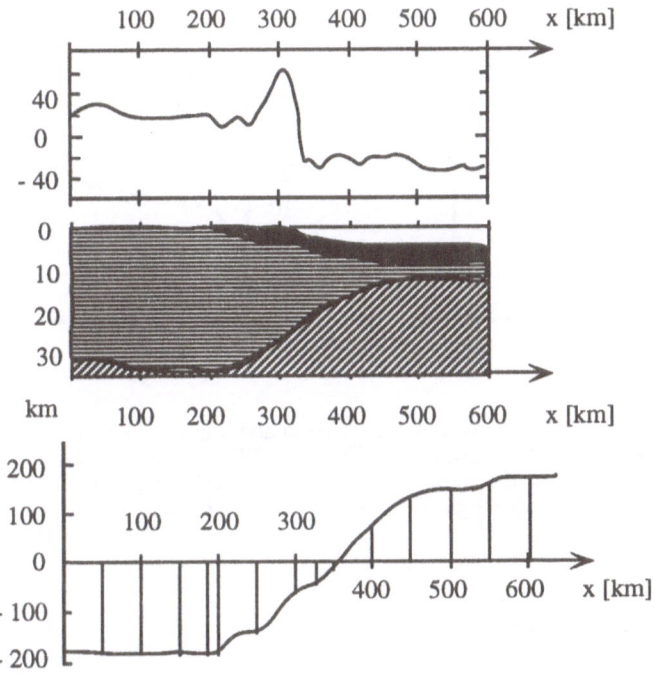

Fig. 3.12. Gravity profile for an SVD inversion example. **Top:** The original Mt. Desert gravity profile (Worzel 1965). **Middle:** Geological cross-section. **Bottom:** The Mt. Desert Bouguer gravity profile with effects of the sea and the sediments are removed (Oldenburg 1974). (from Pedersen 1977)

For his study of SVD Pedersen chose N = 16 data points along the profile. The SVD analysis showed that the eigenvalues from the second to the last, the 16th, have the same order of magnitude. However, experiments gradually increasing the number of eigenvalues showed (Fig. 3.12) a drastic drop in the measure of fit (denoted by ε in Fig. 3.13). The model for 14 eigenvalues (parameters) resembles very much the Bouguer anomaly itself (compare the models in Fig. 3.14 with the profile in Fig. 3.12c). This is typical for a gravity problem, where the topography of a bedrock surface is determined by the assumption that the densities on each side of the boundary remain constant along the whole profile. When increasing the number of parameters from 14 to 16 the model exhibits greater oscillations, but the resolution-error analysis shows an improved resolution for the model boundary in the ranges [x = 150 to 200 km] and [x = 300 to 350 km]. There seem to be variations of the Moho of the order 3-4 km in both ranges.

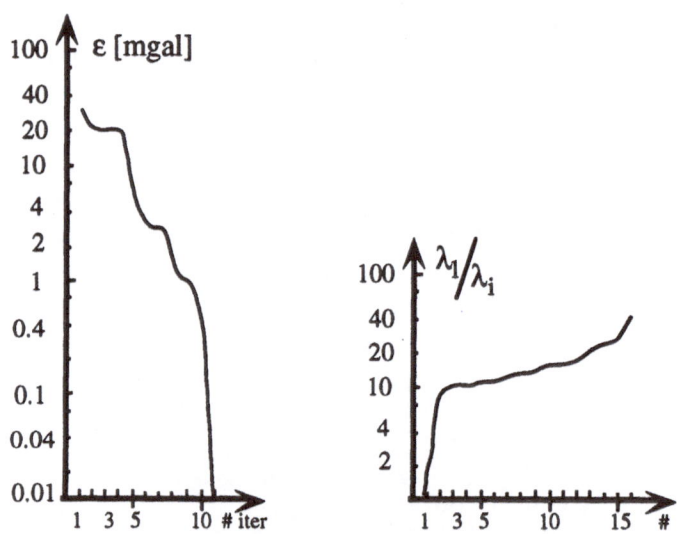

Fig. 3.13. Properties of the iteration and eigenvalues for the Mt. Desert gravity SVD inversion (after Pedersen 1977). **Left:** Decrease in the objective function, the number of eigenvalues has been changed during the iteration process. **Right:** Normalized inverse eigenvalues.

3.4.3.2 Earth Density Model (Weidelt 1978)

Suppose the density of a three-layer Earth is to be determined from known values of the total mass, $M = 6 \cdot 10^{24}$ kg, and inertia, $J = 8 \cdot 10^{37}$ kg m². The layers are numbered 1 to 3, starting with 1 for the core, 2 for the lower mantle and 3 for the combined upper mantle and crust. The radius of the Earth is R = 6371 km and the layer boundaries are located at $R_{12} = a R$ and $R_{23} = b R$ with a = 0.5 and b = 0.75. The model parameters are linear, but instead of the densities ρ_i parameters proportional to the volume of each layer are used:

a.

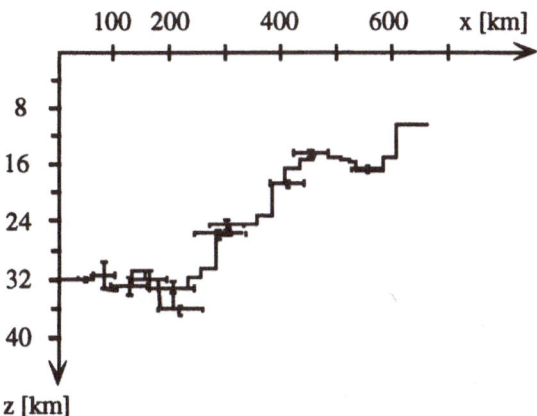

b.

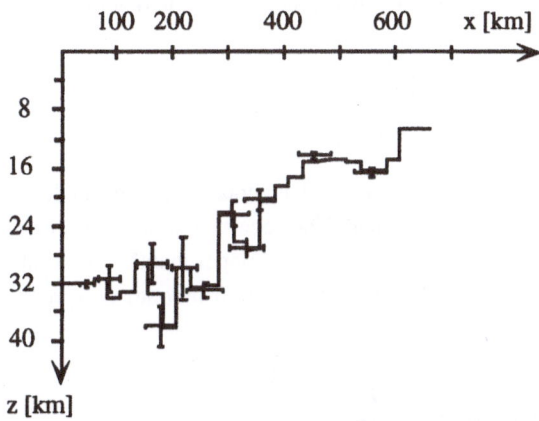

Fig. 3.14. SVD-inverted models and error-resolution bars for the Mt. Worzel gravity data (Pedersen 1977). **a.** (on the previous page) $p = 14$ eigenvalues. **b.** $p = N = M = 16$ eigenvalues.

$$p_1 = a^3 \, \rho_1, \quad p_2 = (b^3 - a^3) \, \rho_2 \quad \text{and} \quad p_3 = (1 - b^3) \, \rho_3.$$

Correspondingly, the data are normalized so that the dimensions of both data and parameters are the same (density):

$$F_1 = M/V = 3M/(4\pi R^3) \quad \text{and} \quad F_2 = J/(2VR^2/5) = 15J/(8\pi R^5).$$

Then the matrix $\underline{G}$ (dimensionless) and the data vector F (in g/cm^3) are:

$$\underline{G} = \begin{pmatrix} 1 & 1 & 1 \\[2mm] a^2 & \dfrac{b^5 - a^5}{b^3 - a^3} & \dfrac{1 - b^5}{1 - b^4} \end{pmatrix} = \begin{pmatrix} 1.000 & 1.000 & 1.000 \\ 0.250 & 0.694 & 1.319 \end{pmatrix}$$

$$\mathbf{F_h} = \begin{pmatrix} 3M/(4\pi R^3) \\ 15J/(8\pi R^5) \end{pmatrix} = \begin{pmatrix} 5.539 \\ 4.549 \end{pmatrix} [\text{g/cm}^3].$$

The eigenvalues are determined from:

$$|\mathbf{G}\,\mathbf{G}^T - \lambda^2\,\mathbf{I}| = \begin{vmatrix} 3.000 - \lambda^2 & 2.263 \\ 2.263 & 2.255 - \lambda^2 \end{vmatrix} = 0,$$

giving $\lambda_1^2 = 4.938$; $\lambda_2^2 = 0.351$ and the data eigenvector matrix $\mathbf{U}$ is:

$$\mathbf{U} = \begin{pmatrix} 0.7603 & -0.6496 \\ 0.6496 & 0.7603 \end{pmatrix}.$$

The parameter eigenvector matrix $\mathbf{V}$ becomes (the problem is underdetermined):

$$\mathbf{V} = \begin{pmatrix} 0.4154 & -0.7757 & 0.4751 \\ 0.5453 & -0.2057 & -0.8127 \\ 0.7281 & 0.5966 & 0.3376 \end{pmatrix}.$$

The first vectors tell us that both data values give about equal information to the densities of the layers, the density of the core slightly weighted. The second vector shows, that most information is gained about the density difference between the core and upper mantle. Least information is available for the density of the lower mantle. This becomes very clear from both the inverse matrix $\mathbf{H}$:

$$\mathbf{H} = \mathbf{V}\,\Delta^{-1}\,\mathbf{U}^T = \begin{pmatrix} 0.9928 & -0.8740 \\ 0.4122 & -0.1046 \\ -0.4050 & 0.9787 \end{pmatrix},$$

and from the resolution matrix $\mathbf{R}$:

$$\mathbf{R} = \mathbf{V}\,\mathbf{V}^T = \begin{pmatrix} 0.7742 & 0.3861 & -0.1603 \\ 0.3861 & 0.3397 & 0.2743 \\ -0.1603 & 0.2743 & 0.8861 \end{pmatrix},$$

whereby the second diagonal element (corresponding to the parameter p_2 of the lower mantle) is by far the smallest of the diagonal elements. The solution finally is:

$$\mathbf{p} = \mathbf{H}\,\mathbf{F_h} = \begin{pmatrix} 1.523 \\ 1.807 \\ 2.209 \end{pmatrix} [\text{g/cm}^3].$$

Transformed into true densities and with additional calculation of variances (in parentheses) in units g/cm^3 gives:

Core: $\rho_1 = 12.18$ (10.58)
Lower mantle: $\rho_2 = 6.09$ (1.43)
Upper mantle and crust: $\rho_3 = 3.82$ (1.83).

Since the second eigenvalue is very much smaller than the first one, it may be interesting to compare some results for a single eigenvalue. The density parameters are not changed significantly, whereas the variances are reduced remarkably. The resolution of all parameters becomes very poor indeed:

For one eigenvalue:

$$\mathbf{H}_1 = \begin{pmatrix} 0.1422 & 0.1215 \\ 0.1866 & 0.1595 \\ 0.2492 & 0.2129 \end{pmatrix},$$

$$\mathbf{R}_1 = \begin{pmatrix} 0.1726 & 0.2265 & 0.3024 \\ 0.2265 & 0.2973 & 0.3970 \\ 0.3024 & 0.3970 & 0.5301 \end{pmatrix}$$

$$\mathbf{P}_1 = \begin{pmatrix} 1.340 \\ 1.759 \\ 2.349 \end{pmatrix} [g/cm^3]$$

Core: $\rho_1 = 10.72$ (1.50)
Lower mantle: $\rho_2 = 5.97$ (0.825)   1 eigenvalue only.
Upper mantle and crust: $\rho_3 = 4.06$ (0.567)

3.4.3.3 Magnetometry (Pedersen 1977)

The paper by Pedersen (1977) concludes with a similar analysis of a magnetic data profiles. The real data example comes from the continental shelf in southwestern Greenland (the data profile is shown in Fig. 3.15 a). The interpretation supposes that the basement is magnetic in contrast to the overlaying non-magnetic sediments and water layers. The problem is then again to determine the topography of the interface, in this case of the top of the basement. With $N = 27$ data points parameters, a reasonable inversion of the magnetic Greenland data was obtained with $M = 18$ or 19 parameters, from which onwards, the eigenvalues decreased almost exponentially (linear increase for the inverse eigenvalues on the logarithmic scale of Fig. 3.15 b). The magnetic problem is more non-linear in nature than the gravity problem explaining the differences in behaviour: less efficient iteration, more critical dependence on the initial model and greater error-resolution sensitivity.

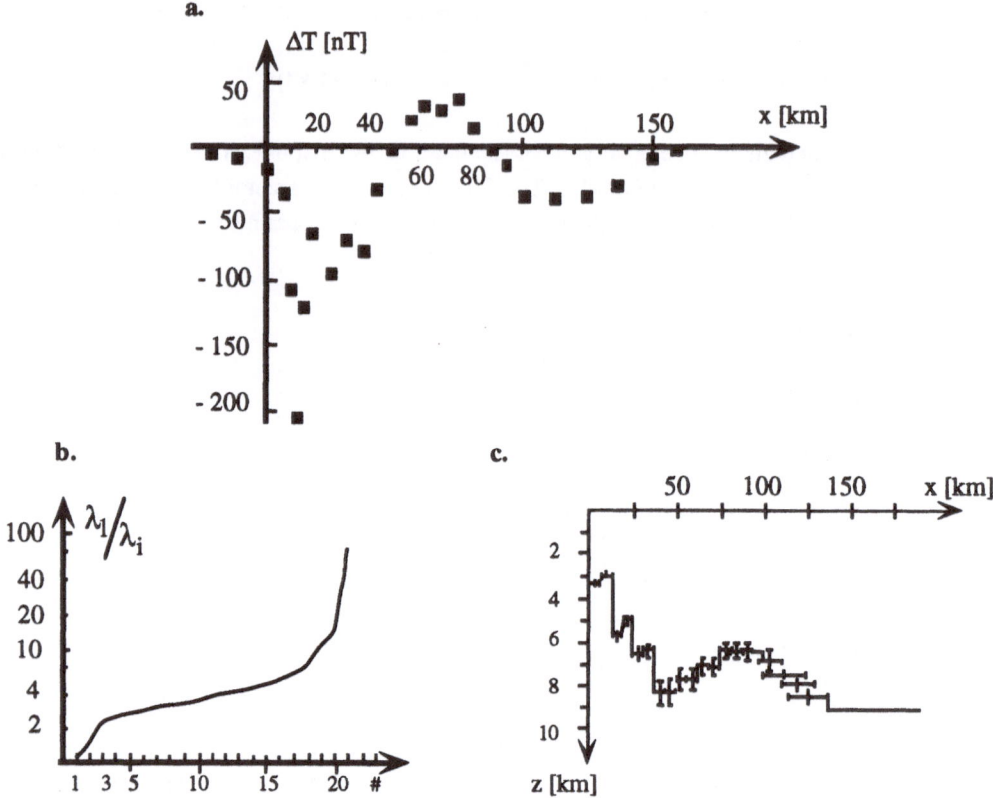

Fig. 3.15. SVD-inversion of a magnetic profile from the continental shelf of SW Greenland (Pedersen 1977). **a.** The data. **b.** Inverse normalized eigenvalues. **c.** Inverted model for 20 eigenvalues

3.4.3.4 Magnetometry, Theoretical Plate Models (Hjelt and Pedersen 1977 unpublished)

SVD has been used to analyze the behaviour of the thick dyke (or plate) model in magnetic inversion (Hjelt and Pedersen unpublished 1977). The same model (although with differing numerical values of the parameters) will be considered later in connection with non-linear optimization. Figures 3.16 - 3.18 show the model, the corresponding anomaly profile, the data eigenvectors **u** and the parameters eigenvectors **v** for three of the models. The plate model is two-dimensional and can be described by six parameters:

$p_1 = d$ = horizontal thickness of the plate;
$p_2 = x_o$ = horizontal location of the plate (the midpoint of its upper surface);
$p_3 = z_o$ = depth of the upper surface of the plate;
$p_4 = h$ = vertical (depth) extent of the plate:
$p_5 = \phi$ = dip of the plate;
$p_6 = k$ = magnetic susceptibility of the plate.

In the first analyzed example, the plate has an infinite depth extent. Since the susceptibility is, self-evidently, the parameter which determines the amplitude of the anomaly, it has not been included in this analysis. Thus $\mathbf{p} = (p_1 = d, p_2 = x_o, p_3 = z_o, p_5 = \phi)$. The result indicates that $\mathbf{v}_1$, associated with the highest and strongly dominant eigenvalue $\lambda_1 = 8.51$, is completely dominated by the dip (Fig. 3.16). The corresponding data eigenvector tells, that the data points along the steepest slope of the anomaly have greatest information content with respect to the dip parameter. These facts are well known for geophysicists having practiced magnetic interpretation. Many "rule of thumb" dip interpretation techniques have been designed over the years which have been based on the slope of the magnetic anomaly. The remaining eigenvalues $\lambda_2 = 0.853$, $\lambda_3 = 0.742$ and $\lambda_4 = 0.293$ and the corresponding parameter eigenvectors give information on the position and size of plate in the order x_o, z_o and d respectively. Except for the susceptibility (which was not analyzed), the order of importance of the parameters differs from the order determined on the basis of correlation analysis in non-linear optimization studies of the magnetized plate model (Hjelt 1973)

The last eigenvalue is dominated by the plate width and the greatest contribution to this parameter comes from points close to the half amplitude of the anomaly. This seems to confirm a commonly use "rule of thumb" connecting the half-width of of the magnetic anomaly to the width of the plate model. In magnetic interpretation practice it is also often claimed that the ratio of depth to width of the plate is a unique parameter, while the parameters depth and the width, separately, are not. The belief is not uniquely confirmed by this analysis. The reason is obviously that the strictness of this rule is dependent on the other parameters (= the form) of the plate.

In Fig. 3.17 the plate has a finite depth extent. The eigenvalues, $\lambda_1 = 135$, $\lambda_2 = 0.94$, $\lambda_3 = 0.67$ and $\lambda_4 = 0.51$ with the first being very strongly dominant. This can be explained as follows: since the dip and the vertical extent of the plate are correlating parameters, the information content of the data is shared by both parameters. In the example of Fig. 3.16, where the vertical extent is infinite, this parameter ceases to influence the anomaly and all the information content of the slopes relates to the dip. The three smallest eigenvalues bear information on the plate parameters in the order z_o, x_o and d, with a slightly different emphasis and order than in the finite depth extent case.

The model in Fig. 3.18 consists of two almost parallel plates, which have a finite, although large depth extent. The model is a modification of inverting real measured data (see Hjelt 1973, 1975) in contrast to the two preceding ones, which were theoretical constructs only. Figure 3.18 shows the eigenvectors for all 12 parameters of the two plates. Now $\mathbf{p} = (p_1 = h_1, p_2 = \phi_1, p_3 = d_1, p_4 = z_{o1}, p_5 = x_{o1}, p_6 = k_1, p_7 = h_2, p_8 = \phi_2, p_9 = d_2, p_{10} = z_{o2}, p_{11} = x_{o2}, p_{12} = k_2)$. As expected, the susceptibilities of the plates are the dominating parameters, they determine the amplitudes of the two anomaly peaks of the profile. The higher peak is caused by the more strongly magnetized plate, the first eigenvalue is dominated by its strongest loading at k_2. The ratio of the loadings k_2 and k_1 depends clearly on the amount of merging (points along the profile, where both plates contribute significantly) between two anomaly peaks. The following parameter in order of importance is, not surprisingly, the dip of the plates. The merging of the partial anomalies complicates the analysis of the order of importance of the other parameters. The most surprising fact (at least in contrast to intuitive thinking) is that the horizontal thicknesses of the plates, the d-parameters, have dominating loadings in the ninth and tenth parameters only. The least important parameters are the depth extents (= vertical thicknesses), h. This is in agreement with expectations, since the lower boundaries of the plates do have a very small influence on the anomaly profile.

The least important parameters are the depth extents (= vertical thicknesses), h. This is in agreement with expectations, since the depth extents of the plates are great. The lower boundaries of the plates do thus have a very small influence on the anomaly profile.

a.

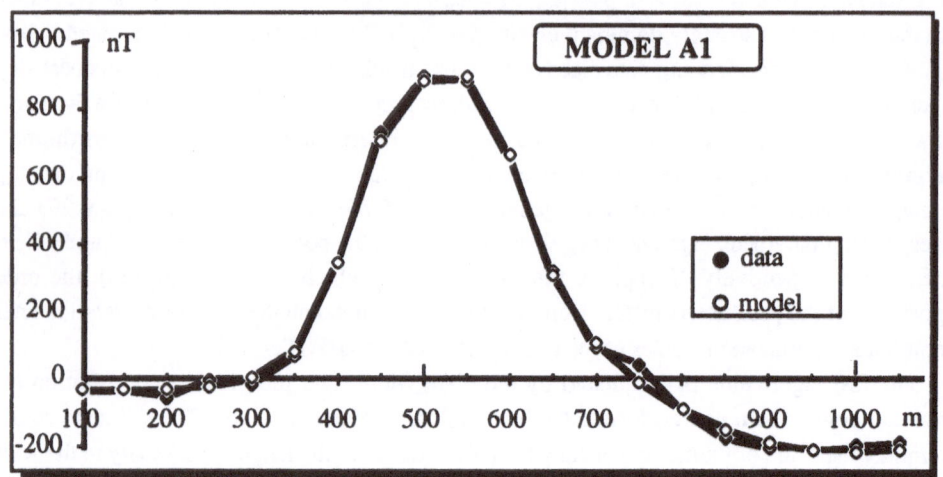

b.

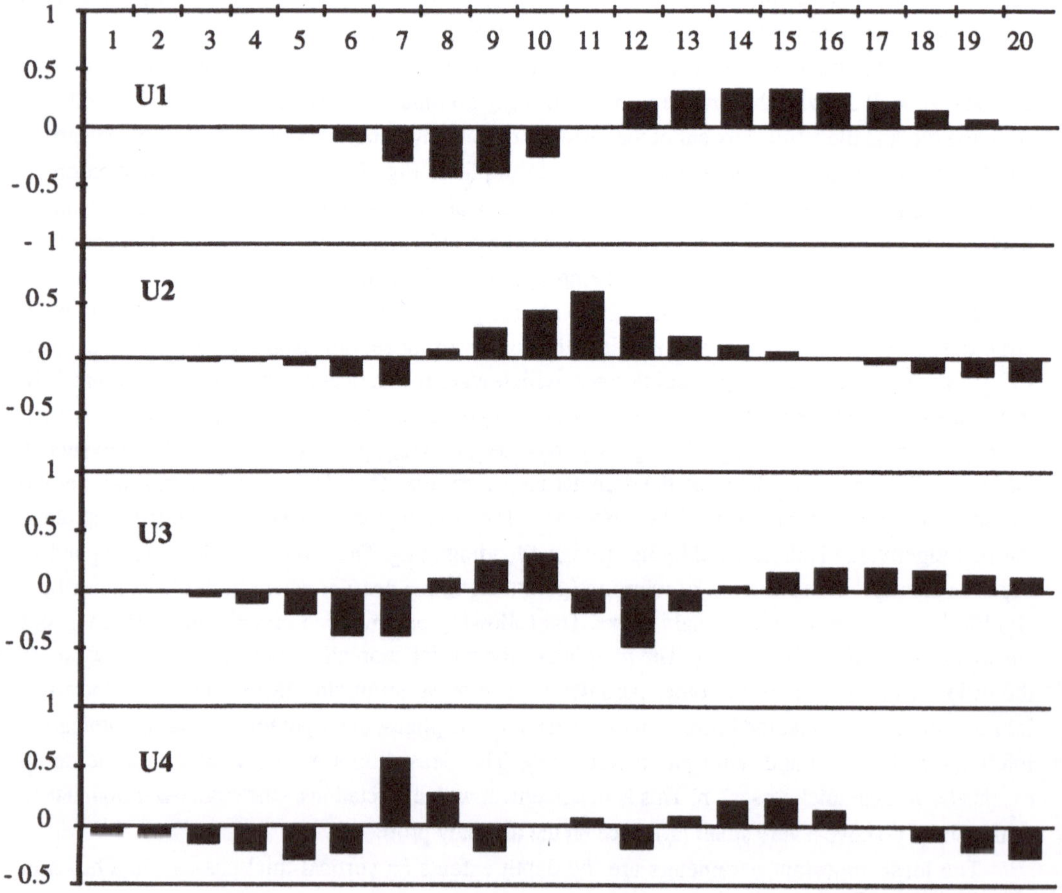

(Fig. 3.16 continues)

(Fig. 3.16 continued)

c. **d.**

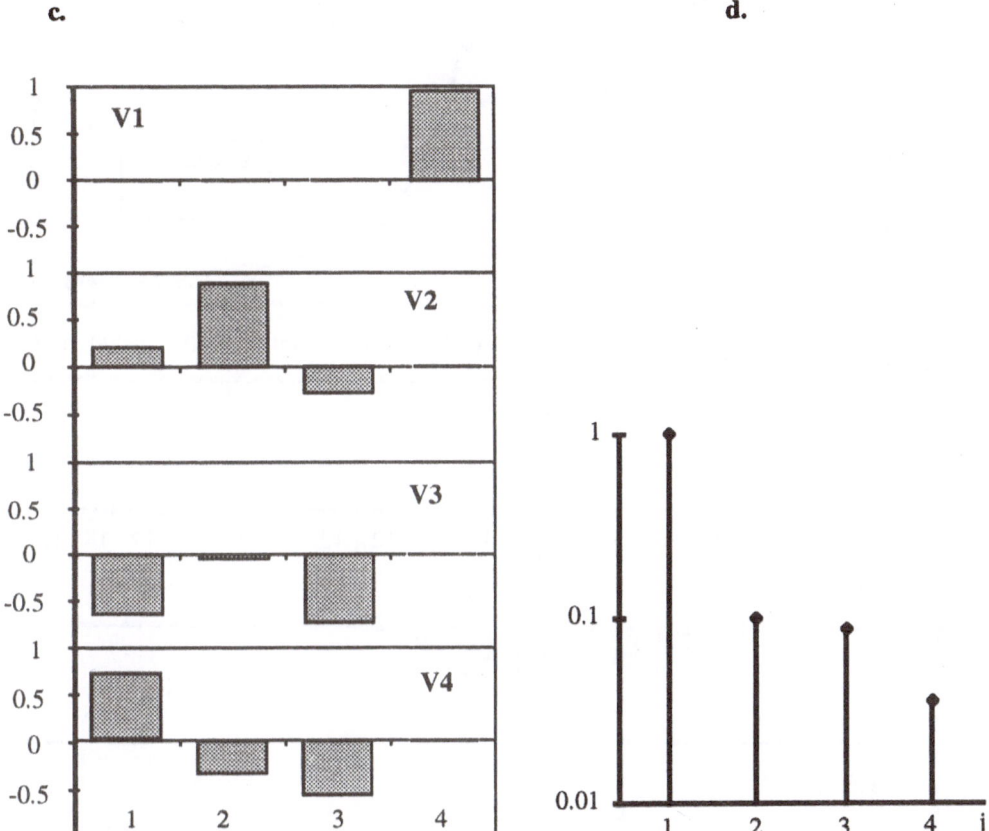

Fig. 3.16. SVD test for magnetized thick plates.   Model A1 (finite depth extent):
d = 200 m, x = 500 m, z = 50 m, h = 250 m (fixed), fii = 30 degrees,
k = 0.01 cgs (fixed). **a.** Data and model profiles, **b.** Data eigenvectors,
c. Parameter (model) eigenvectors, **d.** Normalized eigenvalues.

a.

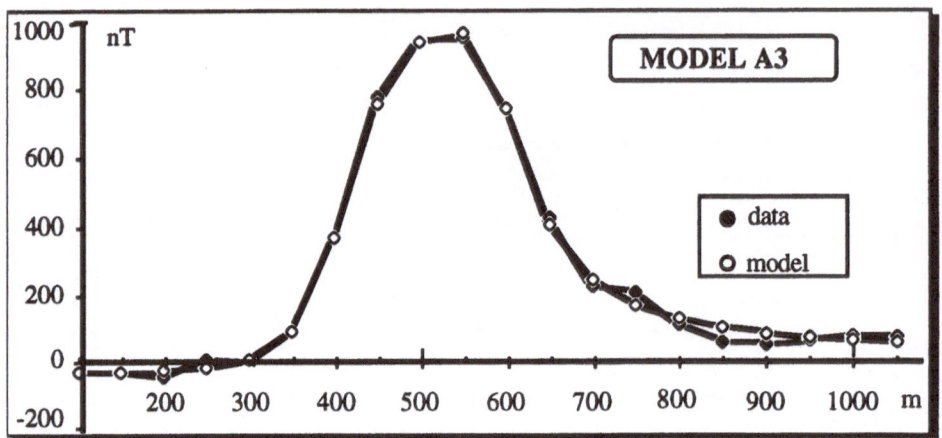

b.

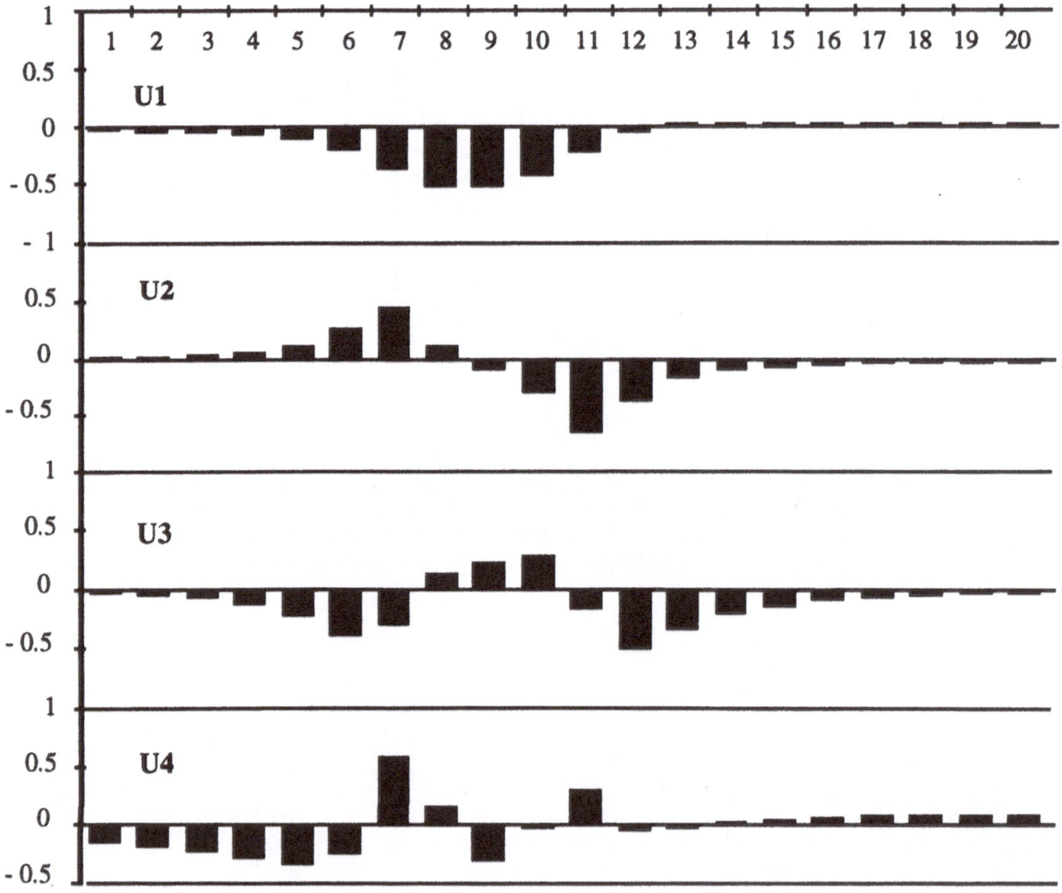

(Fig. 3.17 continues)

(Fig. 3.17 continued)

c.   d.

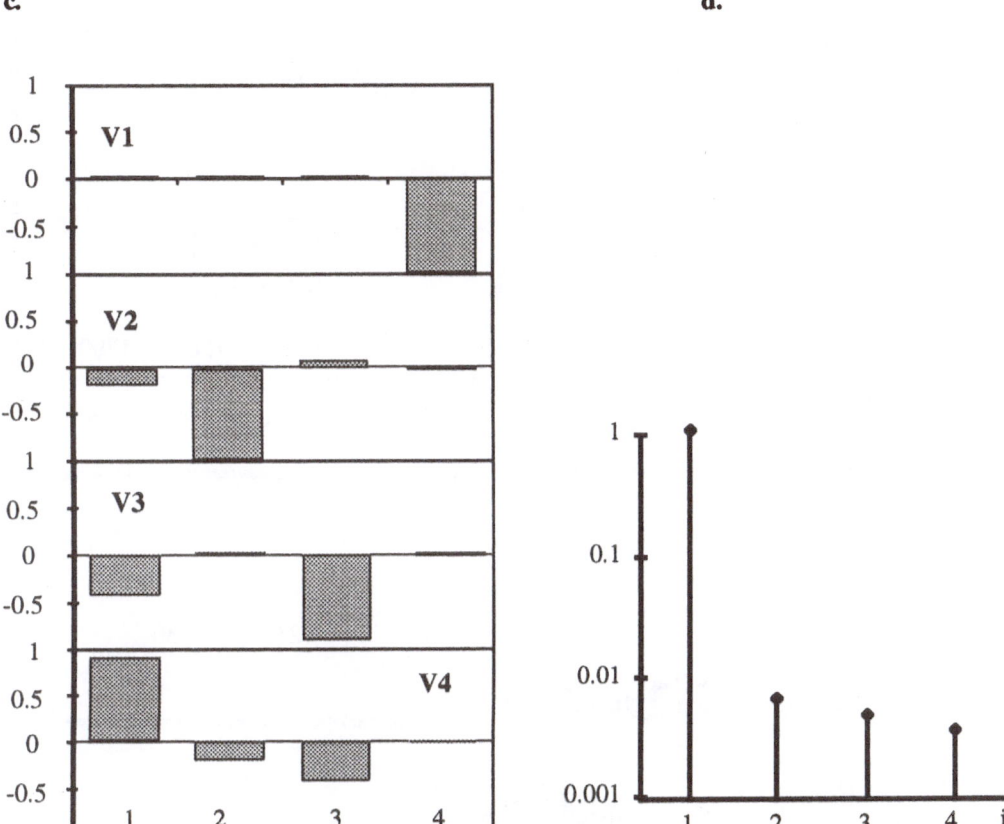

Fig. 3.17. SVD test for magnetized thick plates. Model A3 (infinite depth extent):
d = 200 m, x = 500 m, z = 50 m, h = ∞ (fixed), fii = 30 degrees, k = 0.01 cgs (fixed). **a.** Data and model profiles, **b.** Data eigenvectors, **c.** Parameter (model) eigenvectors, **d.** Normalized eigenvalues.

Fig. 3.18. SVD test for two magnetized thick plates. Model C4 (infinite depth extent):

PLATE 1: d = 200 m, x = 500 m, z = 50 m, h = ∞ (fixed), fii = 75 degrees, k = 0.01 cgs (fixed);
PLATE 2: d = 200 m, x = 800 m, z = 50 m, h = ∞ (fixed), fii = 75 degrees, k = 0.01 cgs (fixed).
a. Data and model profiles. **b.** Data eigenvectors, **c.** Parameter (model) eigenvectors.

(Fig. 3.18 is on pages 88 and 89)

(Fig. 3.18)

a.

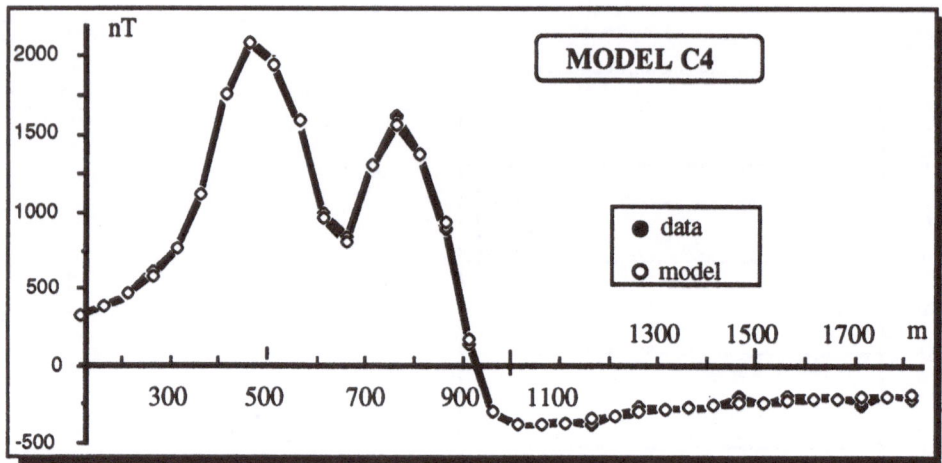

b.

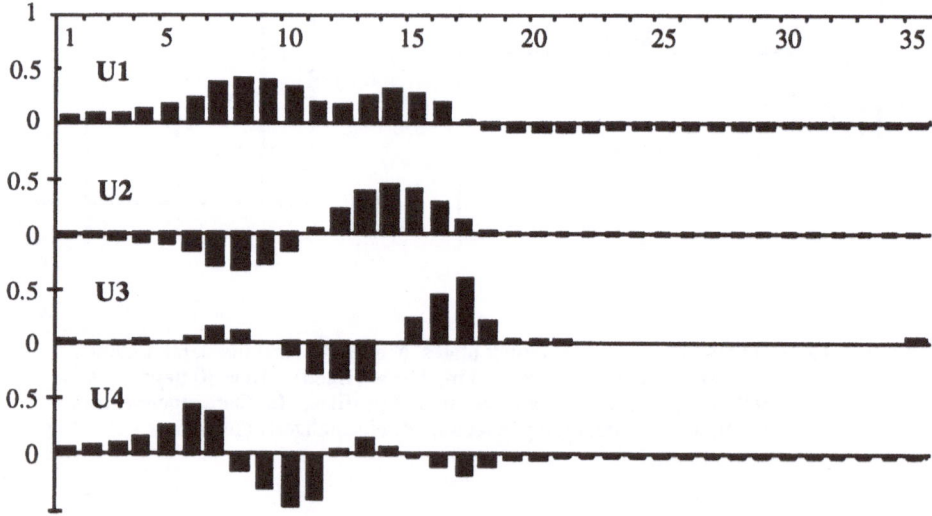

(Fig. 3.18 continues)

(Fig. 3.18 continued)

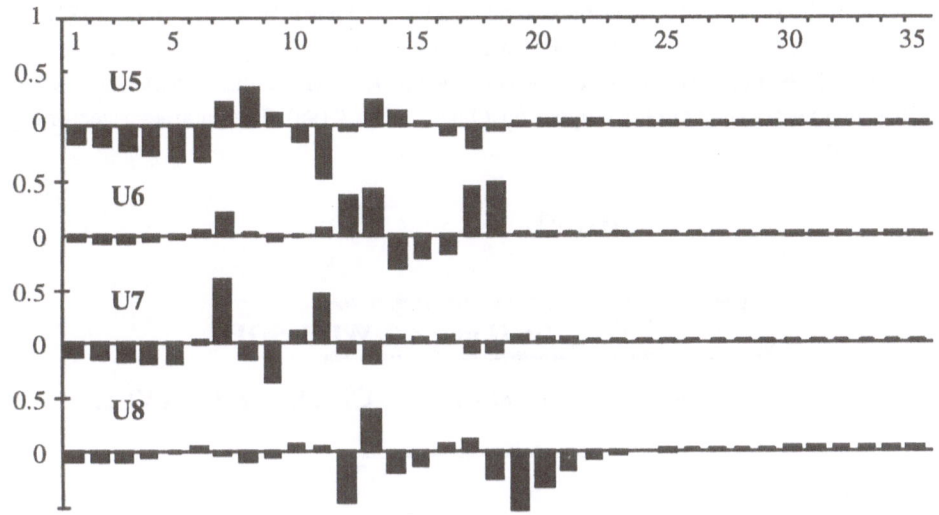

c.

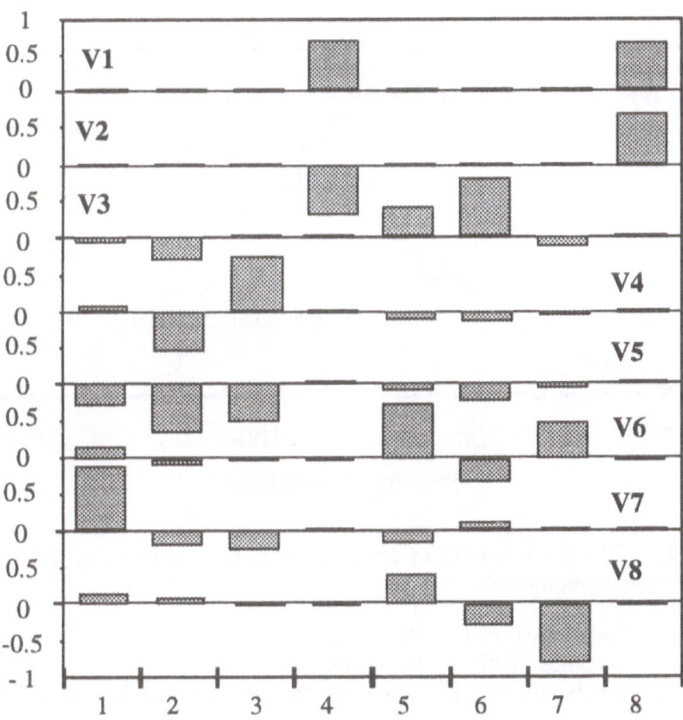

3.4.3.5 Electromagnetism (Inman et al. 1973)

One of the very early articles presenting an SVD analysis is by Inman et al. (1973) for an EM geometrical sounding system. The model was a simple three-layer Earth, wet soil and clay/gravel laying over the bedrock. The model can be described by five parameters: resistivities ρ_i of the layers, and the thicknesses (t_i) of the two uppermost layers (Fig. 3.19a). The parameter vector is $\mathbf{p} = (\rho_1, \rho_2, \rho_3, t_1, t_2)$.

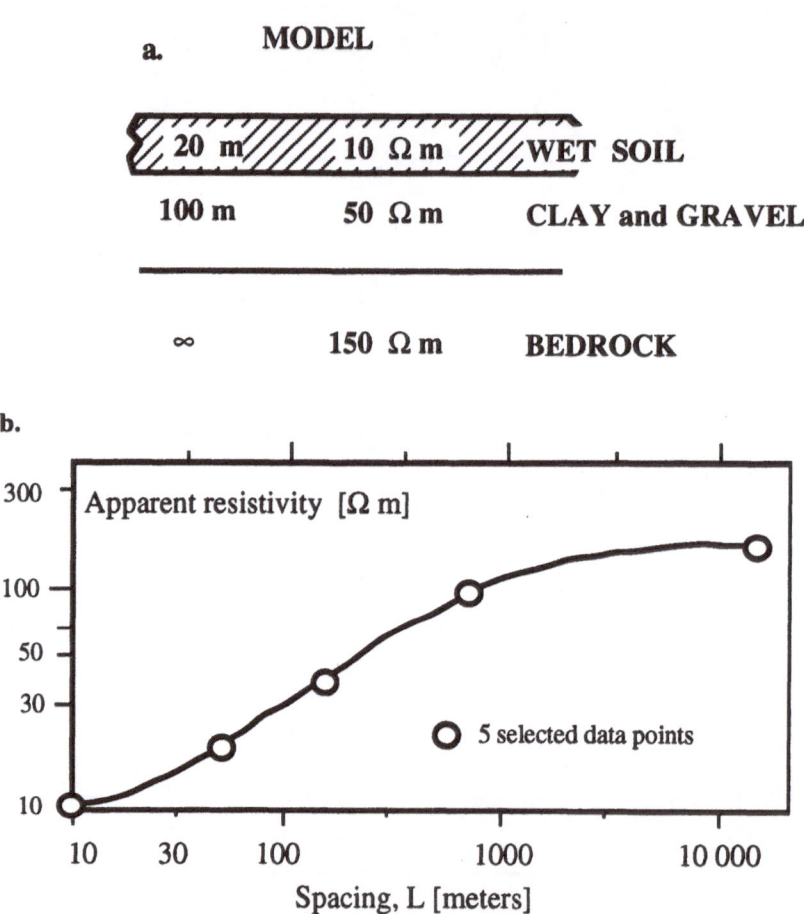

Fig. 3.19. SVD inversion of EM sounding data (After Inman et al. 1973).
a. Three-layer Earth model:

Layer #:	1	2	3
Resistivity [Ωm]:	10	50	150
Thickness [m]:	20	100	∞

b. Apparent resistivity curve, 17 transmitter-receiver spacings, L = 10 .... 10 000 m.

c.

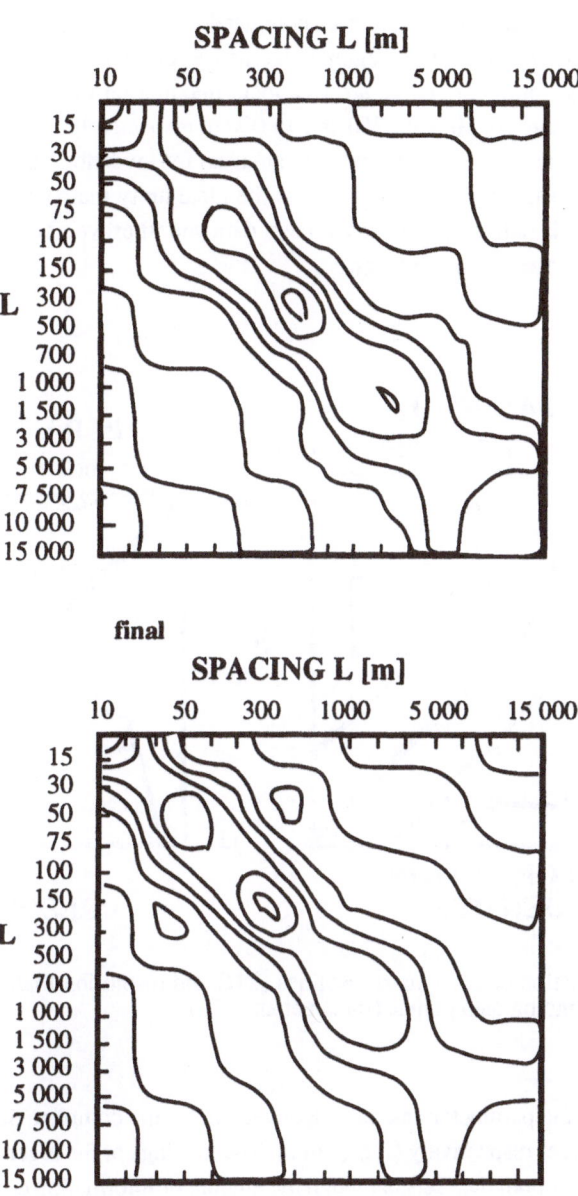

Fig. 3.19c. The information density matrix isolines for the complete sounding data set. The initial model (**top**) and the final model (**bottom**). (Inman et al. 1973).

The measuring system consisted of an EM vertical magnetic dipole as transmitter and a vertical receiver dipole, the sounding effect is produced geometrically, i.e. by measuring the response of the Earth at a number of transmitter-receiver spacings, L. The complete data set consisted of apparent resistivities taken at 17 spacings (five per decade).

The information density matrix was presented both for the initial model of the iterative inversion as well as for the resulting model (Fig. 3.19c). The presentations differ from those shown earlier in that the elements of the matrix have been combined into a equi-density map form. The ideal matrix would be a straight line along the diagonal, the matrix for the final model departs from this, but is clearly diagonally dominant. There are five local maxima on the diagonal of the map, at transmitter-receiver spacings 10, 50, 150, 750 and 15 000 m respectively. The underlined ones are the most distinct ones. If these five data points only are used for inversion, the internal structure of the information matrix disappears and it approaches the ideal unity matrix as does the resolution matrix (Fig. 3.19d). This is rather a paradox: with less data you improve your resolution, naturally by maximizing the effective use of information.

d.

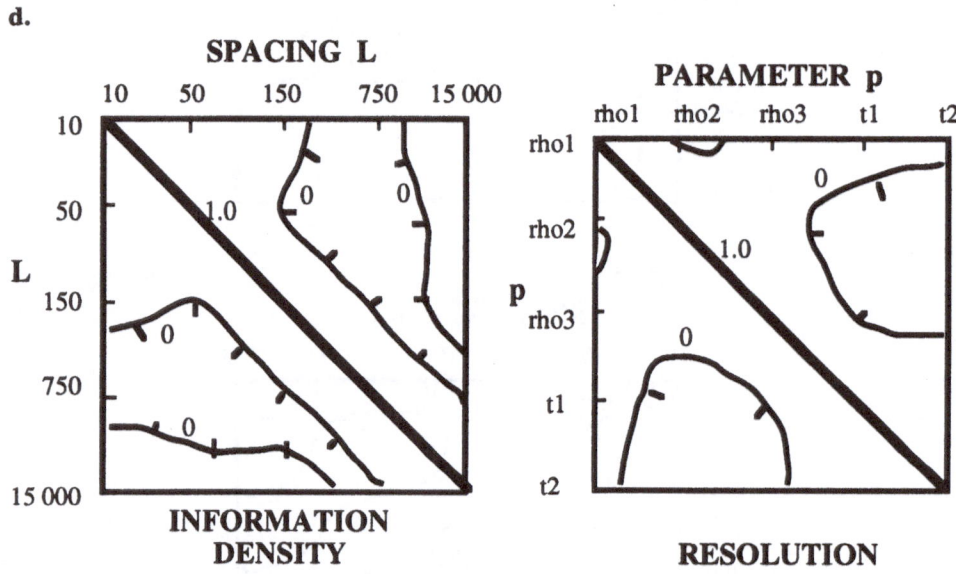

Fig. 3.19d. Information density matrix isolines (**left**) and resolution matrix isolines (**right**) for five selected sounding data points. (Inman et al. 1973).

Fig. 3.20 provides the parameter and data eigenvectors for the complete data set (Fig. 3.20 a) and the five distinct data set respectively (Fig. 3.20 b). The loadings of the eigenvector components are qualitatively similar for both data sets, although the amount of information content of the various data points varies between the two cases. The following discussion concerns the complete data set only. Two parameters belong to the highest eigenvalue $\lambda_1 = 6.86$; the resistivity and thickness of the topmost layer (ρ_1 and t_1). In fact, they have loadings of opposite sign and almost equal size, which means that their ratio ρ_1/t_1 (or its inverse) is more important for the interpretation than the parameters separately. This ratio is actually the inverse conductance of the first layer, receiving its information from the rising part of the apparent resistivity sounding curve, a fact well known to geophysicists working with electrical sounding. The next eigenvalue, $\lambda_2 = 2.03$, depends almost solely on the resistivity of the lowermost layer, the bedrock. The information to this parameter, ρ_3, comes from the large-spacing part of the sounding curve. This is exactly what is to be expected since the original data show, that enough large spacings have been used, so that the apparent resistivity curve levels out at the true resistivity of the bottom layer.

The sounding curve should start, at the shortest transmitter-receiver spacings, at the true resistivity of the top layer. The original data show that the effect of the higher resistivity of the second layer is evident already as an increase at the second data point (L = 15 m). Thus, it may be less surprising to find that the third eigenvector, $\lambda_3 = 0.68$, relates to the resistivities of the two uppermost layers, ρ_1 and ρ_2, but also on their thicknesses t_1 and t_2. The information comes from the very shortest transmitter-receiver spacings, but also from the strongly rising part of the sounding curve. The fourth eigenvalue, $\lambda_4 = 0.25$, connects with the second layer parameters, in fact with its conductance S_2 rather than with the resitivity, ρ_2 and thickness, t_2 individually. Part of the information concerns the thickness of the top layer, t_1. The information comes from a dozen smallest spacings on the sounding curve. The last eigenvalue, $\lambda_5 = 0.062$, carries also information on the conductance of the second layer, S_2, taking its information from most of the points of the sounding curve.

a.

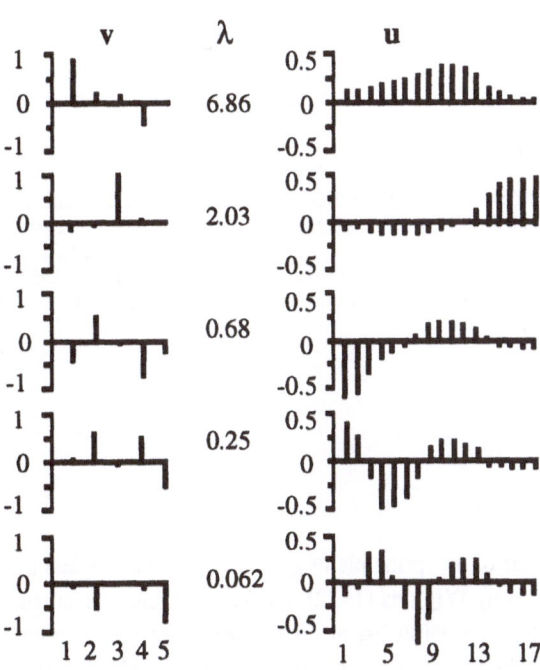

(Fig. 3.20 continues)

(Fig. 3.20 continued)

b.

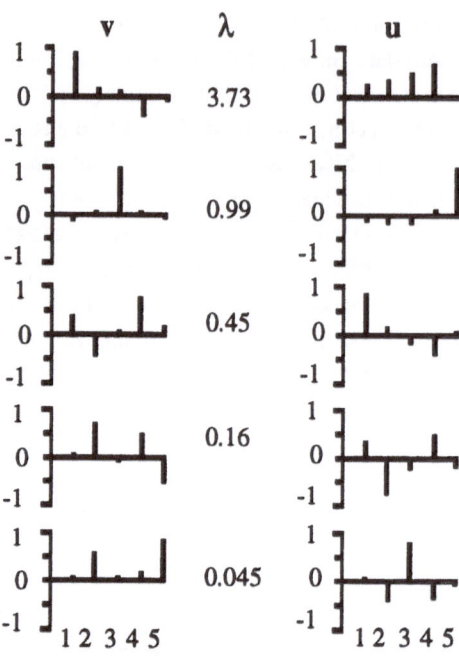

Fig. 3.20. Parameter eigenvectors **v**, the eigenvalues themselves and the data eigenvectors **u** (from left to **right**) for the final model. The components 1 to 5 of **v** correspond to ρ_1, ρ_2, ρ_3, t_1 and t_2 respectively. **a.** All data points. The components 1 to 17 of **u** correspond to L = 10, 15, 30, 50, 75, 100, 150, 300, 500, 7500, 1 000, 1 500, 3 000, 5 000, 7 500, 10 000 and 15 000 m respectively. **b.** For the five selected data points. The components 1 to 5 of **u** correspond to L = 10, 50, 150 and 15 000 m respectively. (after Inman et al. 1973)

3.4.3.6 Seismology (Wiggins 1972).

One of the first applications of the principle of generalized inversion and the SVD technique to geophysical data was published by Wiggins (1972). He analyzed seismic surface waves and Earth free-oscillation data. He suggested using linear combinations of the original linear parameters whenever the solution to the original set of equations is not unique. Various linear combinations of the eigenvectors of the coefficient matrix provide insight into the parameter resolution and the information distribution among the data. This shows how to use the data more effectively, where the information comes from among the data and how the set of possible models is constrained.

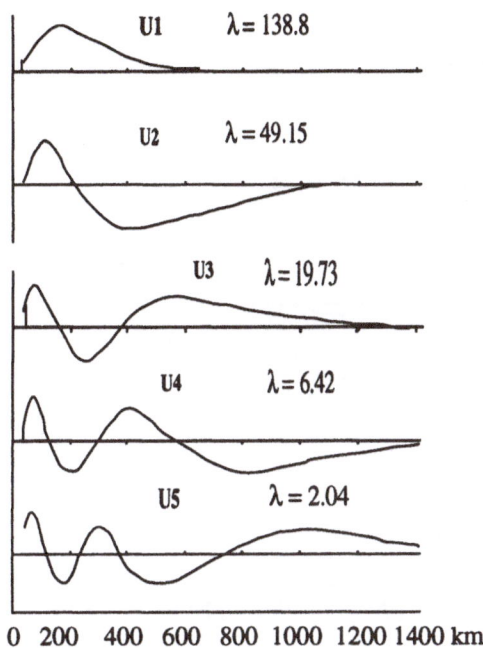

Fig. 3.21. Data eigenvectors of partial derivatives of five selected fundamental spheroidal oscillation modes with respect to shear-wave velocity. (Redrawn from Wiggins 1972).

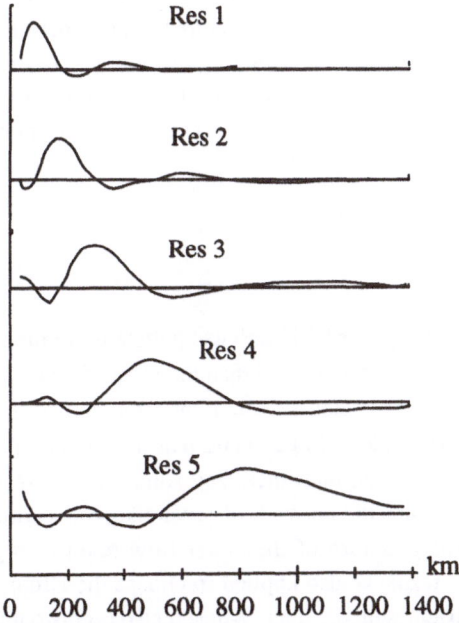

Fig. 3.22. Parameter resolution matrix of the original Earth oscillation data. The order of the "vectors" is the same as for the eigenvectors. (Redrawn from Wiggins 1972).

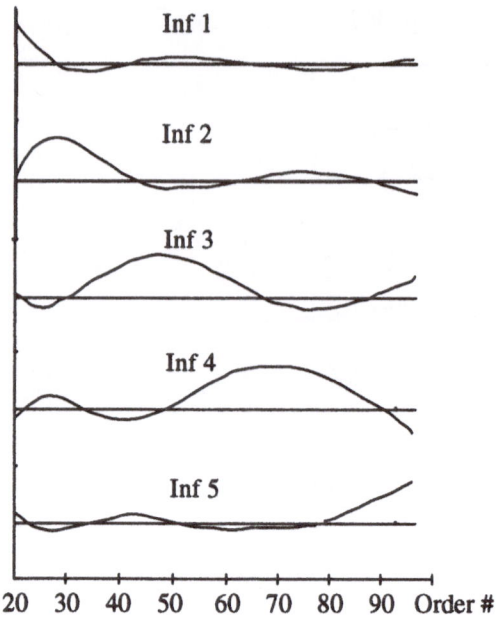

Fig. 3.23. Information density matrix [information distribution according to Wiggins (1972)].

Wiggins concludes in the abstract of his paper: " ... computation of parameter and information resolution is such a simple extension of any inversion procedure based on perturbation parameters that such inversion studies are incomplete without considering resolution." Redrafts of his presentation of the data eigenvectors, the resolution matrix and the information density matrix (Wiggins calls it information distribution) speak for themselves. The presentations in a continuous form are shown in Figs. 3.21 - 3.23.

3.4.3.7 Oceanography

Wunsch (1978) discussed in an early paper SVD and its application to the problem of water circulation in the oceans. The model is more complicated than the others discussed, and in addition the problem is a dynamic one. Wunsch demonstates the use of the method by a strongly simplified model, pure geostrophic flow. The conservation of mass is used as a constraint for the model. The mathematical framework is that of discretized and linearized solution. The resulting system of equations is strongly underdetermined, since the network of oceanographic recording stations was sparse and a proper description of the structure of the water flow requires a great number of parameters. The principle of Occam's razor is also applied to choose the simplest model (of the infinite number of possibilities) consistent with the data. Wunsch also points out the deep analogy between studying the resolution of the problem by SVD techniques and between conventional spectral analysis and filtering theory. The resolution matrix gives hints, where station pairs can be eliminated because no structure is resolved by them and where to add stations in the hope of deducing additional spatial structure.

3.4.3.8 Joint Inversion (Raiche et al. 1985)

The most advanced geophysical inversion techniques combine data obtained by several methods in order to reduce the non-uniqueness of the problem. The combined use of DC and EM sounding has been described in papers by Jupp and Vozoff (1977), Vozoff and Jupp (1975) and Raiche et al. (1985). In the two first papers Schlumberger DCS and magnetotelluric (MT) sounding data for five-layer models were studied. The last paper analyzes a combination of Schlumberger DCS and TEMS (transient EM sounding) data for different three-layer Earth models. The TEM system is using a 100 m x 100 m square loop both as transmitter and receiver, the transient is measured starting with 0.4 ms from the end of the transmitter pulse. The current electrode distances in the DCS is assumed to vary between 1 and 2000 m. The classical three-layer models are labelled A, Q, H and K, where the letter symbols are used to a qualitative description of following resistivity sequences:

A:	resistivity increases with depth
H:	a conducting mid-layer
K:	a resistive mid-layer
Q:	resistivity decreasing with depth

In the following, an H-type example from Raiche et al. (1985) is briefly described. There are two stages of linearization in their inversion approach. Firstly, the problem is linearized by using logarithms of both the data (expressed as apparent resistivities) and the model parameters. Secondly, the inversion is performed iteratively, the linear change of the logarithmic parameters being the unknowns of the linear equations. A three-layer model requires five parameters for a complete description, thus $p = (\rho_1, \rho_2, \rho_3, t_1, t_2)$ {in fact $(\log[\rho_1], \log[\rho_2], \log[\rho_3], \log[t_1], \log[t_2])$}. The linear equation is solved in Raiche et al. (1985) by the modified Marquardt method (see Chap. 4 on non-linear optimization), in which the diagonal elements of the coefficient matrix are modified by a damping factor so that instead of $1/\lambda_i$ the diagonal elements of $[G^{-1}]$ are:

$$[G^{-1}]_{ii} = t_i/\lambda_i \qquad \text{with } t_i = k_i^4/[\mu^4 + k_i^4] \text{ and } k_i = \lambda_i/\lambda_1,$$

where μ is the accepted lower boundary of the eigenvalues. The paper also has tables containing the confidence limits of the parameters assuming 5% Gaussian noise added to the DCS data and 10% Gaussian noise to the TEMS data. The joint inversion works equally well also for the A and K models as analyzed by Raiche et al. (1985).

For the true H-model $p = (100\ \Omega m, 0.25\ \Omega m, 1000\ \Omega m, 30\ m, 10\ m)$ and the apparent resistivity curves are shown in Fig. 3.24. The data show that the DCS curve has a good level to determine the resistivity of the first layer, whereas the maximum AB is too short to determine reliably the level of the bottom layer resistivity. When the DCS data are inverted alone, then ρ_1 and t_1, the parameters of the first layer and thus the location of the upper boundary of the conducting layer

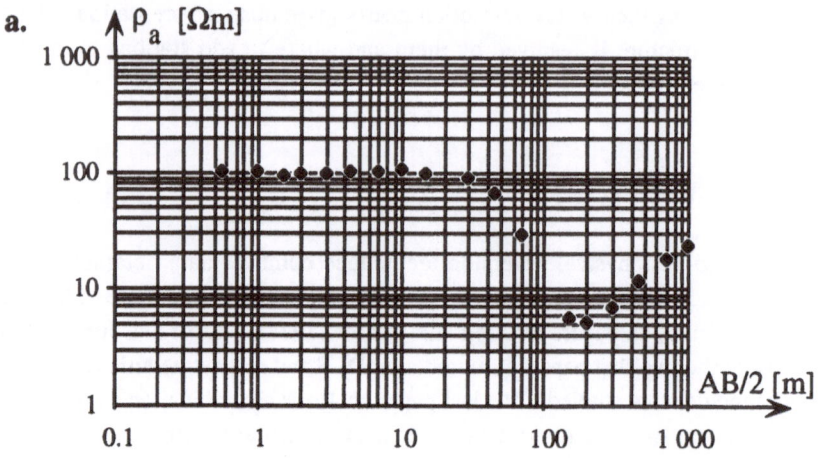

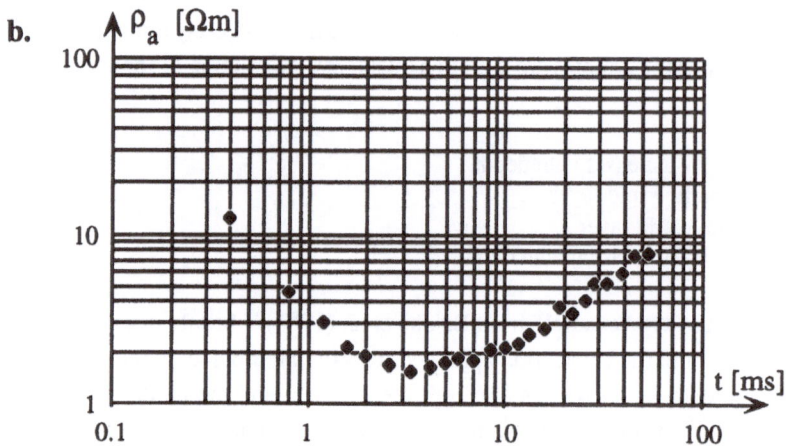

Fig. 3.24. Apparent resistivity curves for the joint inversion example.
a. DCS Schlumberger data. **b.** TEMS concident loop data.
(from Raiche et al 1985).

are reasonably well resolved (Fig. 3.25). The resistivity of the second layer, ρ_2, and its thickness, t_2, are in error by a factor of 4, the resistivity of the bottom layer, ρ_3, is about 30 % too low. However, if one compares the conductance of the second layer for the true model and the DCS inversion, the agreement is fairly good. The inversion of TEMS data gives only the boundary between layers 2 and 3 with reasonable accuracy, all other parameters are more in error than the in the DCS inverted model. The joint inversion gives a good agreement for all parts of the model. Since the accuracy of ρ_3 for TEMS data is much worse than for the DCS inversion, joint inversion cannot improve the accuracy of this parameter from the DCS case.

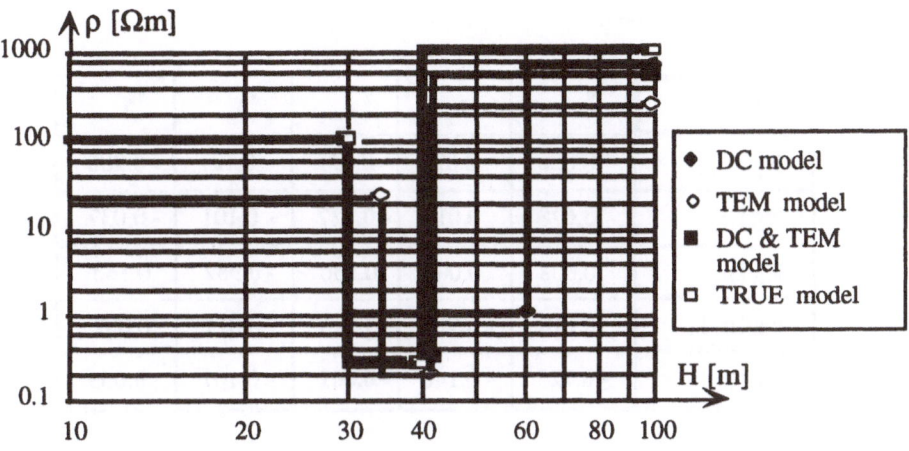

Fig. 3.25. Inversion models for separate and joint inversion as well as the true model.
(after Raiche et al. 1985)

When looking at the parameter eigenvector components (Table 3.2.), for the DCS, TEMS and joint inversion separately, the above conclusions are confirmed. The parameters of the first layer dominate the first eigenvector (v_1) of DCS inversion. The conductance of the second layer is dominant in the eigenvector v_2 and is included also in eigenvectors v_1 and v_3. The conductance of the first layer dominates the eigenvector v_3 and the resistivity of the bottom layer dominates the eigenvector v_4. The product of the second layer parameters is dominant in the eigenvector v_5.

For TEMS inversion the conductance of the second layer dominates the first and third eigenvectors v_1 and v_3. The thickness of the first layer dominates v_2 and the resistivity of the first layer v_5. The resistivity of the bottom third layer is strongly present in the eigenvectors v_4.

TABLE 3.2.   Parameter Eigenvectors (from Raiche et al. 1985). Dominant components are shown in boldface.

DCS	v_1	v_2	v_3	v_4	v_5
log [ρ_1]	0.571	- 0.437	**0.695**	- 0.001	0.001
log [ρ_2]	0.275	**0.629**	0.169	- 0.076	**0.703**
log [ρ_3]	0.004	0.012	0.005	**0.996**	0.094
log [h_1]	**0.726**	- 0.127	**- 0.676**	0.003	- 0.008
log [h_2]	- 0.267	**- 0.630**	- 0.177	- 0.057	**0.705**

(Table 3.2 continues)

(Table 3.2 continued)

TEMS	v_1	v_2	v_3	v_4	v_5
$\log [\rho_1]$	0.026	- 0.023	0.091	- 0.121	**- 0.988**
$\log [\rho_2]$	**0.678**	0.015	**- 0.727**	- 0.101	- 0.037
$\log [\rho_3]$	0.008	0.006	0.136	**- 0.982**	0.133
$\log [h_1]$	0.118	**- 0.989**	0.086	0.011	0.032
$\log [h_2]$	**- 0.725**	- 0.148	**- 0.661**	- 0.107	- 0.063

JOINT	v_1	v_2	v_3	v_4	v_5
$\log [\rho_1]$	0.012	- 0.351	**0.936**	- 0.033	- 0.000
$\log [\rho_2]$	**0.684**	0.035	- 0.031	**- 0.729**	- 0.013
$\log [\rho_3]$	0.004	0.002	0.002	0.022	**- 1.000**
$\log [h_1]$	0.135	**- 0.924**	**- 0.345**	0.092	0.000
$\log [h_2]$	**- 0.717**	- 0.144	- 0.070	**- 0.678**	- 0.018

For the joint inversion the eigenvectors are very clearly structured and logical, the order of decreasing accuracy of the mentioned parameters and parameter combinations is clearly evident:

v_1: conductance of second layer
v_2: thickness of first layer
v_3: resistivity of first layer
v_4: resistivity-thickness product of second layer
v_5: resistivity of third layer

3.4.4 Ridge Regression

A variant of the Singular Value Decomposition (SVD) method has been designed to address the question of singularities of the coefficient matrix $\underline{G}$ (or $\underline{G}^T \underline{G}$). In this ridge regression method, an additional parameter k is introduced, so that:

$$P_{RR} = \frac{\lambda_j}{\lambda_j + k} P_{LSQ} \qquad \text{for } j = 1, 2, ..., M. \qquad (3.55)$$

The subscripts RR and LSQ stand for ridge regression and least squares respectively and M is the number of eigenvalues of the multiplying matrix. Alternatively different additional parameters k_j can be for different eigenvalues. The effect of ridge regression is to emphasize the least-squares solution, when the eigenvalues are much greater than k (or k_j). When the eigenvalues $\lambda_j \ll k_j$, then the corrected parameters will be much smaller than the least-squares ones.

The inverted parameters are given by:

$$P_{RR} = (\mathbf{G}^T\mathbf{G} + \mathbf{Q})^{-1}\,\mathbf{G}^T\mathbf{F}_h. \tag{3.56}$$

If $\mathbf{U}$ is the eigenvector matrix of $\mathbf{G}^T\mathbf{G}$, the additional matrix $\mathbf{Q}$ can be written using the decomposition

$$\mathbf{Q} = \mathbf{U}\,\mathbf{K}\,\mathbf{U}^T \tag{3.57}$$

$$\text{with } \mathbf{K} = \begin{pmatrix} k_1 & & & 0 \\ & k_2 & & \\ & & \ddots & \\ 0 & & & k_M \end{pmatrix}$$

The effect of the procedure is seen to be closely related with and very similar to the Levenberg-Morrison-Marquardt method of non-linear optimization. For a further study a good description of the ridge regression method can be found in Weisberg (1985).

3.5   THE BACKUS-GILBERT APPROACH

3.5.1   Definitions

One of the first systematic and mathematical descriptions of the geophysical inversion problem is in a series of publications by Backus and Gilbert (1967, 1968, 1970). They soon had, together with the newly introduced SVD technique, a major impact on research of geophysical inversion especially but also on mathematical inversion in general. A series of elegant papers introduced these new ideas to the geoscience community (Backus 1970a, b, c; Franklin 1970; Parker 1972; Wiggins 1972 and many, many others). New ideas and systematic introduction of probabilistic approaches were soon to reach the English-speaking geophysical world (Sabatier 1974, 1977, 1978, 1979; Tarantola and Valette 1982a, b; Tarantola 1987, just to mention a few). This brief introduction to the principles of Backus & Gilbert follows the description of Nolet (1981).

Backus and Gilbert specifically worked with continuous parameter functions, which depended on one coordinate only. In their case the parameter was the density of the Earth as a function of depth (radius). Backus and Gilbert did show that a group of models, which fit a finite set of data,

$$F_{hi} = \int G_i(r) \, p(r) \, dr \qquad\qquad (3.58)$$

is either empty or has an infinite dimension (= ∞ many models fit the data). Above F_{hi} are the data, G the kernel function of the theoretical (model) field and p(r) the model parameter function. The formulation does not require any further knowledge about the model parameters.

The two main sources of uncertainty of the model are:
1. Finite (insufficient) amount of data
2. Data errors.

Backus and Gilbert searched for an answer to the question: which of the possible models should be chosen? They specified that the norm of the parameter function (= its distance from a given or assumed basic model) should be as "small" as possible. The norm of a function in Euclidean space is:

$$\| p(r) \| = \int_0^R p^2(r) \, dr \,, \qquad\qquad (3.59)$$

leading to the minimization problem:

$$Q = \| p(r) \| - \sum_j^M 2 \, \lambda_j \int G_j(r) \, p(r) \, dr \, = min! \,, \qquad\qquad (3.60)$$

where λ_j are Lagrangian multipliers. In other words:

$$\int_0^R [\, 2p - 2\sum_M \lambda_j G_j] \, \partial p \, dr = 0. \qquad\qquad (3.61)$$

So $p = \sum_j^M \lambda_j \, G_j$ and the solution is:

$$F_{hi} = \sum T_{ij} \, \lambda_j \qquad\qquad (3.62)$$

$$\text{with } T_{ij} = \int_0^R G_i(r) \, G_j(r) \, dr \,.$$

The problem with this solution is, that for many geophysical problems, the kernel function G is not quadratically integrable.

Backus and Gilbert define the local average of the model parameter as:

$$< p(r_0) > = \int_0^R A(r,r_0) \, p(r) \, dr \qquad\qquad \int_0^R A(r,r_0) \, dr = 1 \,. \qquad\qquad (3.63)$$

The width of the region for calculating the average, R, describes the resolution of the model parameter. The resolution is best, when $A(r, r_0) = \partial(r - r_0)$, $<p(r_0)> = p(r_0)$ or in other words the inversion gives the true value of the parameter. It is neceassry to make a compromise between resolution and the uncertainty (the variance) of the model (which Backus and Gilbert called parameter trade-off, see Fig. 3.26).

3.5.2 Parameter Trade-Off

For any model p(z) fitting the data, the expected parameter average:

$$p_{ave} = \int_0^\infty D_j(z) \cdot p(z) \, dz \qquad (3.64a)$$

is approximately the same. A linear combination of the Fréchet kernels is the Backus-Gilbert averaging function or resolving kernel:

$$A(z, z_0) = \Sigma_i a_n(z_0) D_i(z) . \qquad (3.65)$$

Choosing the coefficients $a_n(z_0)$ in such a way, that the expceted value of the parameter at point z_0:

$$p_{ave}(z_0) = \int_0^\infty A_j(z, z_0) \cdot p(z) \, dz \qquad (3.64b)$$

represents a meaningful average of the parameter p(z).

Ideally one should have $A(z, z_0) = \delta(z - z_0)$. Usually one tries to approach the ideal by minimizing the spread $s(z_0)$ using a quadratic criterion:

$$s(z_0) = 12 \int_0^\infty (z - z_0)^2 \cdot A^2(z, z_0) \, dz \qquad (3.66)$$

together with a constraint:

$$\int_0^\infty A(z, z_0) \, dz = 1 .$$

The factor 12 is determined so that for a pulse function of width w and with an amplitude 1/w, s would be equal to w. Lagrange multipliers are the obvious choice to solve the constrained minimization of s.

If the averaging function is centred at z_0 and has a small spread $s(z_0)$, then $p_{ave}(z_0)$ is a good estimate of the parameter value at z_0. In practice, minimum spread results in a great variance of the corresponding parameter value. If the data are uncorrelated with individual variances ε_n^2, then the variance of the average $p_{ave}(z_0)$ is:

$$\varepsilon^2 = \mathrm{Var}[p_{ave}(z_o)] = \sum_n a_n^2(z_o) \cdot \varepsilon_n^2.$$ (3.67)

The complementary behaviour of variance and resolution is called **trade-off** (see Fig. 3.26). Sometimes it is useful to define an additional function:

$$S(r_o) \cos \beta + \varepsilon(r_o) \sin \beta,$$ (3.68)

which is minimized with respect to ß in order to obtain an optimal balance between the parameter resolution and the variance.

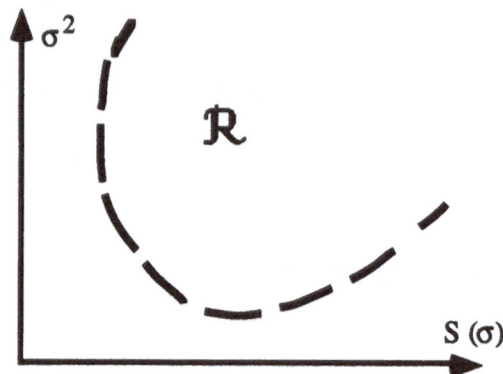

Fig. 3.26. The complementary behaviour of parameter resolution and variance (parameter trade-off).

3.6 GEOPHYSICAL TOMOGRAPHY

Tomography is a technique to study the internal structure of a medium without direct contact with the medium. Energy is transmitted through the region from all sides and measured in as many positions as possible. The first successful and perhaps best-known application of tomography has been in the field of medicine. The signal to be measured has to cross the body, the properties and/or form of which is to be determined. The essential feature of tomographic studies is linearity (or linearization) of the model parameters.

In recent years tomographic inversion has found wide application also in geophysics. There exist already a wide variety of books on the subject. Therefore, only the basic principles and a simple example are considered here. For further information on the subject the reader is referred to the references.

In geophysics it is not always so easy to generate the proper measuring configuration. Two typical measuring constellations have been reported:

1. The region under investigation is covered by two or more boreholes, a source on the surface and measurements are taken in the holes, sometimes also on the surface. Alternatively the source is in one (or several) of the holes and its field is penetrating the body/bodies in question. The measurements are made on the surface and/or in the other boreholes. Most applications have been reported for seismic borehole measurements, but also borehole radar and resistivity tomography have proven to be successful. In principle the radio-wave shadowing method lends itself to tomographic inversion.

2. The seismograph network around the globe is used continuously to register and to monitor signals from earthquakes occurring all over the world. Since the seismic waves propagate through the whole Earth, combining the registrations from a multitude of earthquakes provides genuine tomographic data. Both P and S wave first arrivals have been used in global seismic tomography. The amount of data in this case is so large that supercomputers have been necessary before whole Earth tomographic studies have been possible.

3.6.1 Seismic Tomography

The travel time of seismic energy in homogeneous medium between two points separated by a distance d_j (Fig. 3.27) is:

$$t_j = d_j / v_j. \tag{3.69}$$

where v_j is the velocity of propagation of the seismic waves. If the velocity varies along the seismic wave path, the travel time integral and its digitized version needs to be considered:

$$t = \int_{ray\ path} ds / v(x,y) \rightarrow \sum_{j}^{N_{cell}} d_j / v_j .$$

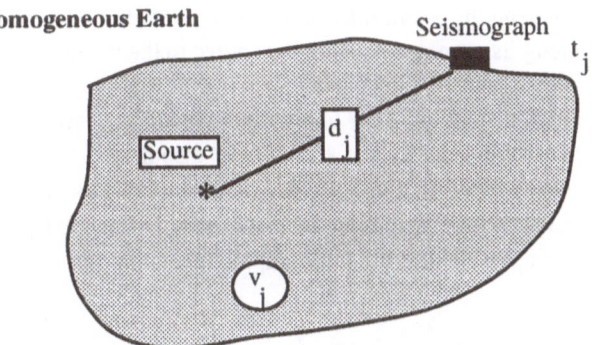

Homogeneous Earth   Seismograph

Fig. 3.27. The geometry of a seismic experiment. The seismic wave travels from the source in an homogeneous Earth (natural: earthquake or artificial: explosion, vibrator etc.) at constant velocity v and is detected at the seismograph from the first arrival time t onwards

In practice the region under investigation is covered with cells, the velocity is assumed to be constant within each cell (Fig. 3.28). Thus the true traveltime becomes a sum of travel times along the whole seismic wave path from the source (the earthquake) to the receiver (the seismometer of the monitoring station):

$$t_{tot} = \sum_{j}^{N} t_j = \sum_{j}^{N} d_j / v_j . \qquad (3.70a)$$

The traveltimes of a great number of seismic ray paths forms the data set and the velocities of each cell is one component of the parameter vector $\mathbf{p} = (v_j); j = 1, 2, ... N_{cell}$. In order to have good variance of the parameters, as many ray paths as possible should cross each of the cells (Fig. 3.29). This calls for relatively large cells resulting in a poor resolution system. The system of equations becomes strongly overdetermined (cf. Fig. 3.30). Decreasing the size of the cells means poorer coverage of rays within each cell and an increase of the variance of the cell parameters. The proper balance is again a question of trade-off between variance and resolution. In addition, the increase in the number of cells increases the size of the system of equations and calls for greater computer capacity.

There are two different ways to introduce the linearity essential for traditional tomographic re-construction. The most straightforward would be to use the inverse of the velocity (a parameter called slowness, s_j, in seismic literature):

$$s_j = 1/v_j$$

The basic tomographic equation then would be

$$t = \underline{D} s, \qquad (3.70b)$$

where t is the vector of travel times of each individual seismic ray, $\underline{D}$ is the matrix of travel paths and s the slowness vector. The components of s are the unknown model parameters to be determined. Each of the path components d_{ij} subdivides into subpaths within those cells the corresponding seismic ray traverses along its propagation from the source to the receiver. The paths are calculable from geometrical considerations.

If one aims at a high resolution in the tomographic inversion, the number of required cells is large and the change in seismic velocity and accordingly, also in slowness from cell to cell, may be small. It is therefore more effective to use another method of linearization, where small changes around some average velocities are used as model parameters. Following the definitions of Aki et al. (1977), one of the first papers on seismic tomographic inversion, the travel time equation can be written as:

$$t_{tot} = t_o + \sum_{j}^{N} t_j \approx t_o + \sum_{j}^{N} (1 + m_j) \, \bar{d}_j / \bar{v}_j , \qquad (3.70c)$$

with $v_j = \bar{v}_j /(1 + m_j)$ and $d_j = \bar{d}_j + \Delta d_j$. The effect of the variations in travel paths within each cell, Δd_j, are assumed to be negligibly small. The small changes in velocity from cell to cell is described by the parameter m_j, actually the fluctuation of slowness. The expressions above would change slightly, if the velocity fluctuations would be used in a similar way.

After some rearrangements one can write the linearized travel time difference:

$$\Delta t = \sum_j^N g_j \, m_j \,, \qquad\qquad (3.71a)$$

with $\Delta t = t_{tot} - t_o - \bar{t}$, $\bar{t} = d_j / \bar{v}_j$ and $g_j = \bar{d}_j / \bar{v}_j$. Taking all travel paths and travel times into account the tomographic equation would become:

$$\Delta t = \underline{G} \, m \,, \qquad\qquad (3.71b)$$

which can be solved with any suitable method available for large systems of linear equations. Aki et al. (1977) used the SVD technique. Some recent improvements of the seismic tomography method consists of taking advantage of the sparseness of the tomography matrix (see Carrion 1990).

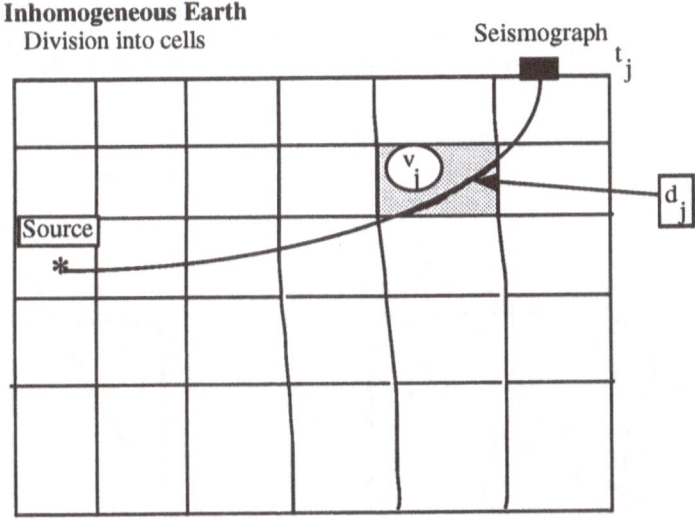

Fig. 3.28. Division of the inhomogeneous Earth into cells of constant velocity.

3.6.2 Tomography and the Radon Transform

The construction problem consists of determining the internal structure of an object without having to cut, crack or otherwise macroscopically damage it. The tomographic technique has been used in a variety of surroundings, the probing energy being mainly EM waves or particles: X-rays, gamma rays, light, microwaves, sound waves, electrons, protons, neutrons and heavy ions have all

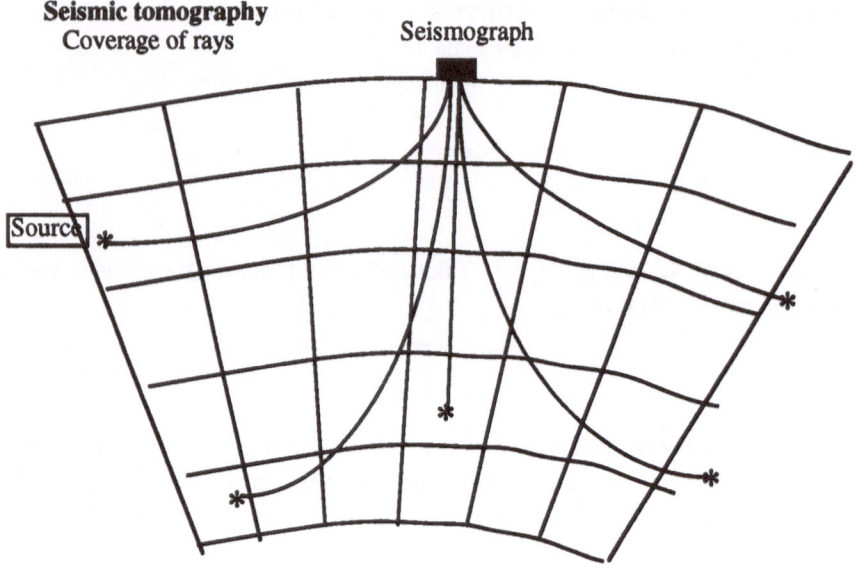

Fig. 3.29. Seismic wave (ray) paths from numerous sources converging on each seismograph are necessary to produce optimum coverage of each velocity cell in seismic tomography.

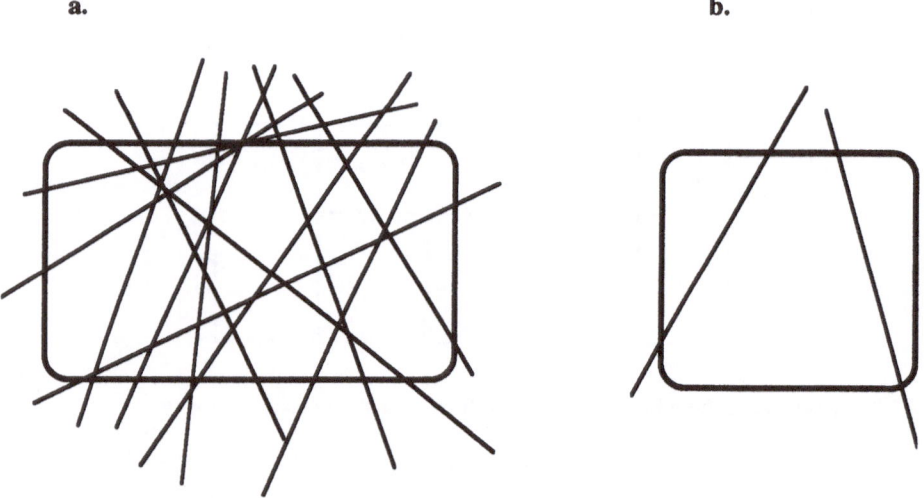

Fig. 3.30. Coverage of seismic rays. **a.** Good coverage: cell parameters are strongly overdetermined and their variance is good. **b.** Poor coverage: Cell parameters are poorly or even underdetermined

been used. In transmission tomography the source of the energy is located outside the volume to be studied, and is typically produced by a man-made controllable source. In emission tomography the (usually natural) physical processes within the volume to be studied are used. Earthquake tomography would fall in the latter category.

The interaction of the probing energy with the distribution of the internal properties of an object can be written in the form of an integral along the line of propagation of the waves or the particles. This can be put in a mathematical frame by using an integral expression known as the Radon transform (RT), R. The detection, the measurement of the energy transmitted, refracted and/or reflected from the interior of the body, $\hat{f}$, consists of one sample of a the Radon transform. The data are called a projected distribution or a profile (Deans 1983):

$$\hat{f} = R(f) .$$
(3.72)

Following the definitions of Deans (1983) and using the notations of this work {given in brackets} we have the chain:

distribution of internal properties {p} -> Action {R} -> Detection {F_h},

which leads to:

$$F_h = R(p).$$
(3.73)

The problem of determining an approximation of p from the measurements F_h was already solved by Radon (1917).

$$p = R^{-1}(F_h) = (-1/\pi) \int_0^\infty \frac{dF_h(q)}{q}$$
(3.74)

This result was forgotten for a long time and in many tomographic applications in fact the relationships were "reinvented" by many authors (Deans 1983). The inverse Radon transform (IRT) is mathematically valid, if p is a continuous function with compact support and if all possible profiles (or projected distributions) F_h are known. This inverse problem is ill-posed and therefore its solution is not unique. The realizations of the inverse RT in tomographic methods boil essentially down to linearized equations. Special techniques have often to be applied to solve for the resulting system of linear equations, since the coefficient matrices are large, but sparse.

There is an intimate connection between the RT and other integral transforms, most notably the Fourier transform (e.g. Deans 1983). In the filtered backprojection approximation for the IRT, the inversion algorithms can be written as (Deans 1983):

$$p = R\,F^{-1}\{|k|\,F\,[(F_h)]\}$$
back projection of a filtered projection

or
(3.75)

$$p = F_2^{-1}\{|k|\,F_2[R\,(F_h)]\}$$
filtering of a back projection.

Many other approaches to IRT exist, among the most popular ones being the iterative methods like ART (= Algebraic Reconstruction Technique) and generalized inverse matrix methods. Further aspects of the Radon transform can be found in Deans (1983) and Natterer (1986).

One can safely say that in a majority of geophysical tomographic applications, the problem has been written in some other mathematical form than the RT. Some of the early exceptions were papers by Chapman (1978, 1979) dealing with the use of RT for the direct problem of seismology. Beylkin (1987) has demonstrated the connection between the Radon transform and the Kirchhoff-type migration method of seismic inversion. On the other hand, tomography and migration respond differently to the low and high wave number parts of the seismic signal as discussed in detail by Mora (1987). Geophysical tomography has recently been covered broadly in two excellent volumes edited by Desaubies et al. (1990) and Nolet (1987) respectively. Both books contain a voluminous list of references, some of the early papers being Lager and Lytle (1976), Aki et al. (1977), Dines and Lytle (1979), Lytle and Dines (1980) and McMechan and Ottolini (1980).

3.6.3   Example of Seismic Regional Tomography (Aki et al. 1977)

The data used in this first exposition of seismic tomography consisted of first arrivals from earthquakes around the globe as registered by the NORSAR seismic array in southern Norway. The epicentre location of the earthquakes used for the analysis was reasonably dense. The spread of the array is not, however, very large, so the tomographic image of the subsurface is limited to a region less than 200 km in diameter. One could call this an example of regional (or even local) tomography. Because of the small size of the region, the curvature of the Earth was neglected in the construction of the tomography equations (but not of course for the main part of the seismic ray paths). The sub-surface was divided into five layers and the average seismic P-wave velocity was constant within each layer. The data for the average model of the region applied by Aki et al. (1977) are given in Table 3.3. Fig. 3.31 is a reproduction of the generalized inverse solution of the uppermost layer, the numbers being slowness fluctuations in percent. The greatest fluctuations amount to 12 %, but this value (as the other encircled block) is not well resolved. The models of velocity variations within the other layers are similar and are therefore not discussed further.

With the improvements in computer capacity, especially that of supercomputers, more demanding tomographic reconstructions have been attempted. Some of the most impressive models can be found in Dziewonski and Woodhouse (1987). The tomographic models are truly global, depicting impressive details in the topography of the mantle-core boundaries and lateral variations of seismic velocities within the mantle.

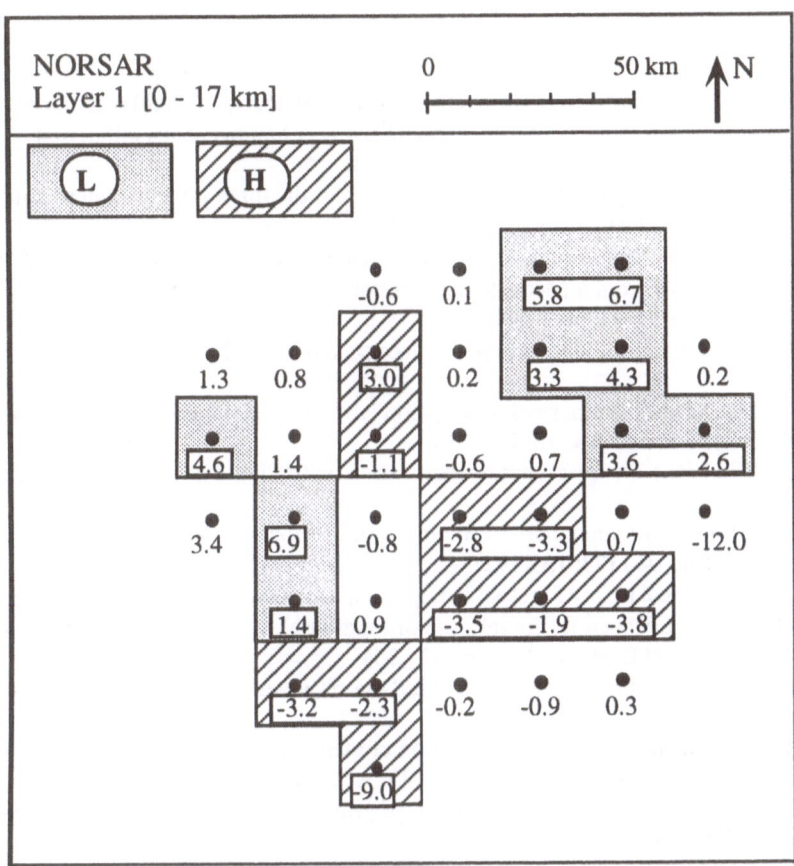

Fig. 3.31. The generalized inverse solution for the first layer (upper crust, 0 - 17 km) of the regional tomography example from Aki et al. (1977). **Numbers** are velocity perturbations in percent of the average layer velocity (6.1 km/s). **L** and **H** are low- and high-velocity anomalies respectively. In the **shaded areas**, the perturbations exceed the standard error

TABLE 3.3. Layered Earth model used in the seismic tomography example (Aki et al. 1977)

LAYER	Thickness (km)	P-wave velocity (km/s)	Location	(km)
1	17	6.1	Upper crust	0 - 17
2	19	6.9	Lower crust	17 - 36
3	30	8.2	Upper mantle	36 - 66
4	30	8.2	Upper mantle	66 - 96
5	30	8.2	Upper mantle	96 - 126

REFERENCES

Aki K, Lee WHK (1976) Determination of three-dimensional velocity anomalies under a seismic array using first P arrival times from local earthquakes 1. A homogeneous initial model. J Geophys Res 81: 4381 - 4399

Aki K, Christoffersson A, Husebye E (1977) Determination of the three-dimensional seismic structure of the lithosphere. J Geophys Res 82: 277 - 296

Backus GE (1970a) Inference from inadequate and inaccurate data I. Proc Nat Acad Sci USA 65:1-7

Backus GE (1970b) Inference from inadequate and inaccurate data II. Proc Nat Acad Sci USA 65: 281 - 287

Backus GE (1970c) Inference from inadequate and inaccurate data III. Proc Nat Acad Sci USA 67: 282 - 289.

Backus GE, Gilbert JF (1967) Numerical application of a formalism for geophysical inverse problems. Geophys J R Astr Soc 13: 247-276

Backus GE, Gilbert JF (1968) The resolving power of gross earth data systems. Geophys J R Astr Soc 16: 169-205

Backus GE, Gilbert JF (1970) Uniqueness in the inversion of inaccurate gross earth data. Philos Trans R Soc Lond 266, A 1173: 123-192

Beylkin G (1987) Mathematical theory for seismic migration and spatial resolution. In: Bernabini M, Carrion P, Jacovitti G, Rocca F, Treitel S, Worthington MH, (eds): Deconvolution and inversion. Blackwell Scient Publ, London, pp 291 - 304

Bjerhammer A (1973) Theory of errors and generalized matrix inverses. Elsevier, Amsterdam, 420 pp

Bott MHP (1973) Inverse methods in the interpretation of magnetic and gravity anomalies. In: Bolt BA (ed): Methods in computational physics, vol 13, Academic Press, USA, pp 123-162

Braile LW, Keller GR, Peeples WJ (1974) Inversion of gravity data for two-dimensional density distributions. J Geophys Res 79: 2017-2021

Burkhard N, Jackson DD (1976) Application of stabilized linear inverse theory to gravity data. J Geophys Res 81: 1513-1518

Carrion PM (1990) Fast dual tomography. Geophys Res Lett 17: 1505 - 1507

Carrion PM, Sampaio EES (1990) Imaging the earth - the quest for resolution. Geophysics: The leading edge of exploration. 9 (10): 30 - 32

Chapman CH (1978) A new method for computing synthetic seismograms. Geophys J R Astr Soc 54: 481 - 518

Chapman CH (1979) On impulsive wave propagation in a spherically symmetric Earth. Geophys J R Astr Soc 58: 229 - 234

Cooke DA, Schneider WA (1983) Generalized linear inversion of reflection seismic data. Geophysics 48: 665 - 676

Deans SR (1983) The Radon transform and some of its applications. John Wiley and Sons, USA, 289 pp

Desaubies Y, Tarantola A, Zinn-Justin J (eds) (1990) Oceanographic and geophysical tomography. NATO Adv Study Inst, Session L, 9.8. - 3.9. 1988. North-Holland, The Netherlands, 463 pp

Dines KA, Lytle RJ (1979) Computerized geophysical tomography. Proc IEEE 67: 1067 - 1073; 1679

Dziewonski AM, Woodhouse JH (1987) Global images of the Earth's interior. Science 236: 37 - 48

Franklin JN (1970) Well-posed stochastic extensions of ill-posed linear problems. J Math Anal Appl 31: 682-716

Glenn WE, Ryu J, Ward SH, Peeples WJ, Phillips JR (1973) The inversion of vertical magnetic dipole sounding data. Geophysics 38: 1109 - 1129

Hartmann RH, Teskey DJ, Friedberg JL (1971) A system for rapid digital aeromagnetic interpretation. Geophysics 36: 891-918

Hjelt S-E (1973) Experiences with automatic magnetic interpretation using the thick plate model. Geophys Prosp 21: 243 - 265

Hjelt S-E (1975) Performance comparison of non-linear optimization methods applied to interpretation in magnetic prospecting. Geophysica 13: 143 - 166

Hjelt S-E (1976) On the use of the Gauss-Seidel method on linear parameter interpretation in magnetometry and gravimetry. Geoexploration 14: 21 - 36

Hohmann GW, Raiche AP (1988) Inversion of controlled source electromagnetic data. In: Nabighian MN (ed) Electromagnetic methods in applied geophysics, vol 1, Theory. Soc Explor Geophys, Tulsa, USA

Inman JR (1975) Resistivity inversion with ridge regression. Geophysics 40: 798 - 817

Inman JR, Ryu J, Ward SH (1973) Resistivity inversion. Geophysics 38: 1088 - 1108

Ivansson S (1987) Crosshole transmission tomography. In: Nolet G (ed) Seismic tomography. (With applications in global seismology and exploration geophysics). D. Reidel, The Netherlands, pp 159 -188

Jackson DD (1972) Interpretation of inaccurate, insufficient and inconsistent data. Geophys J R Astr Soc 28: 97 - 109

Jackson DD (1973) Marginal solutions to quasi-linear inverse problems in geophysics: the Edgehog method. Geophys J R Astr Soc 35: 121 - 136

Jackson DD (1976) Most squares inversion. J Geophys Res 81: 1027 - 1030

Jackson DD (1979) The use of a priori data to resolve non-uniqueness in linear inversion. Geophys J R Astr Soc 57: 137 - 157

Johansen H-K (1977) A man/computer interpretation system for resistivity soundings over a horizontally stratified earth. Geophys Prosp 25: 667-691

Jupp DLB, Vozoff K (1977) Stable iterative methods for the inversion of geophysical data. Geophys J R Astr Soc 42: 957 - 976

Lager D, Lytle RJ (1976) Determining a subsurface electromagnetic profile from high-frequency measurements by applying reconstruction-technique algorithms. Radio Sci 12: 249 - 260

Lanzcos C (1961) Linear differential operators. Van Nostrand, London & Beccles, 564 pp

Lytle RJ, Dines KA (1980) Iterative ray-tracing between boreholes for undergorund image reconstruction. IEEE Trans Geosci Remote Sensing 18: 234 - 240

McMechan GA, Ottolini R (1980) Direct observation of a p - τ curve in a slant stacked wave field. Bull Seismol Soc Am 70: 775 - 789

Mora P (1987) Elastic wavefield inversion for low and high wavenumbers of the P- and S-wave velocities, a possible solution. In: Bernabini M, Carrion P, Jacovitti G, Rocca F, Treitel S, Worthington MH, (eds): Deconvolution and inversion. Blackwell, London, pp 321 - 337

Mora P (1990) A unifying view of inversion. In: Desaubies Y, Tarantola A, Zinn-Justin J (eds): Oceanographic and geophysical tomography. NATO Adv Study Inst, Session L, 9.8. - 3.9. 1988. North-Holland, The Netherlands, pp 345 - 374

Natterer F (1986) The mathematics of tomography. John Wiley and Sons, Great Britain, 222 pp

Nolet G (1981) Linearized inversion of (teleseismic) data. In: Cassinis R (ed): The solution of the inverse problem on geophysical interpretation. Plenum Press, New York, pp 9 - 37

Nolet G (ed.) (1987) Seismic tomography. (With applications in global seismology and exploration geophysics). D. Reidel, The Netherlands, 386 pp

Nolet G, Snieder R (1990) Solving large linear inverse problems by projection. Geophys J Int 103: 565 - 568

Oldenburg DW (1974) The inversion and interpretation of gravity anomalies. Geophysics 39: 526 - 536

Parker RL (1972) Inverse theory with grossly inadequate data. Geophys J R Astr Soc 29: 123 - 138

Parker RL (1977a) Linear inference and unparameterized models. Rev Geophys Space Phys 15: 446 - 456

Parker RL (1977b) Understanding inverse theory. Ann Rev Earth Planet Sci 5: 35 - 64

Pedersen LB (1977) Interpretation of potential field data. A generalized inverse approach. Geophys Prosp 25: 199 - 230

Pedersen LB (1979) Constrained inversion of potential field data. Geophys Prosp 27: 726 - 748

Radon J (1917) Über die Bestimmung von Funktionen durch ihre Integralwerte längs gewisser Mannigfaltigkeiten. Ber Sächsische Akad Wiss (Leipzig) Math-Phys Kl 69: 262 - 267 [English translation in Appendix A of Deans (1983)]

Raiche AP, Jupp DLB, Rutter H, Vozoff K (1985) The joint use of coincident loop electromagnetic and Schlumberger sounding to resolve layered structures. Geophysics 50: 1618 - 1627

Rocca F, Treitel S, Worthington MH (eds): Deconvolution and inversion. Blackwell, London, 355 pp

Rogers PG, Edwards SA, Young JA, Downey M (1987) Geotomography for the delineation of coal seam structure. Geoexploration 24: 301 - 328

Russell B, Hampson D, Chun J (1990) Noise elimination and the Radon transform. Geophysics: the leading edge of exploration. 9 (10): 18 - 23

Rüter H (1987) Migration and tomography: Radon migration. First Break 5: 399 - 402

Sabatier PC (1974) Remarks on approximate methods in geophysical inverse problems. Proc R Soc Lond A 337: 49 - 71

Sabatier PC (1977) On geophysical inverse problems and constraints. J Geophys 43: 115 - 137

Sabatier PC (ed) (1978) Applied inverse problems. Lectures presented at "Etude Interdisciplinaire des Problemes Inverses", Lecture Notes in Physics 85, Springer, Berlin-Heidelberg-New York, 425 pp

Sabatier PC (1979) Comment on the use of a priori data to resolve non-uniqueness in linear inversion by D.D. Jackson. Geophys J R Astr Soc 58: 523 - 524

Spakman W, Nolet G (1988) Imaging algorithms, accuracy and resolution in delay time tomography. In: Vlaar NJ, Nolet G, Wortzel MJR, Cloetingh SAPL (eds): Mathematical geophysics. D. Reidel, The Netherlands, pp 155 - 187

Sørensen K (1976) Førelaesningar i generalized inverse teori. (Lectures in generalized inverse theory, in Danish) Lecture Notes, Univ Aarhus, 46 pp

Tarantola A (1987) Inverse problem theory. Methods for data fitting and model parameter estimation. Elsevier, The Netherlands, 613 pp

Tarantola A, Valette B (1982a) Inverse problems = quest for information. J Geophys 50: 159 - 170

Tarantola A, Valette B (1982b) Generalized non linear inverse problems solved using the least squares criterion. Rev Geophys Space Phys 20: 219 - 232

Vozoff K, Jupp DLB (1975) Joint inversion of geophysical data. Geophys J R Astr Soc 42: 977 - 991

Ward SH, Peeples J, Ryu J (1973) Analysis of geoelectromagnetic data. In: Bolt BA (ed): Methods of Computational Physics, vol. 13, Academic Press, New York, pp 163 - 238

Weidelt P (1978) Einführung in die verallgemeinerte Matrixinversion. Unpubl lecture notes, 27 pp

Weisberg S (1985) Applied linear regression. John Wiley and Sons, New York, 324 pp

Werner S (1955) Interpretation of magnetic anomalies of sheet-like bodies. SGU, Ser C, No 5

Wiggins RA (1972) The general linear inverse problem: implication of surface waves and free oscillations for earth structure. Rev Geophys Space Phys 10: 251-285

Woodhouse JH, Dziewonski AM (1984) Mapping the upper mantle: three-dimensional modeling of Earth structure by inversion of seismic waveforms. J Geophys Res 89, B7: 5953 - 5986

Worthington MH (1984) An introduction to geophysical tomography. First Break, 2 (11): 20 - 26

Worzel JL (1965) Pendulum gravity measurements at sea 1936 - 1959. John Wiley and Sons, New York

Wunsch C (1978) The North Atlantic general circulation west of 50° W determined by inverse methods. Rev Geophys Space Phys 16: 583 - 620

NON-LINEAR PARAMETERS

" To abandon a large pit without knowing where to start digging a new one

is unfair and requires too much from a pragmatic human person . It is

even difficult enough, when the place of a new pit has been selected."

(freely paraphrased from E. de Bono: The Use of Lateral Thinking, 1967)

CHAPTER 4

NON-LINEAR PARAMETERS

4.1 DEFINITIONS

Many geophysical fields depend in a non-linear way on the model parameters, this is especially true for the parameters describing the geometrical features of the models. In the chapter on seismic tomography the material parameter, seismic velocity, in fact appeared in a non-linear way. Although most non-linear problems can be linearized by some of the methods described in Chapter 3, it is instructive to make a survey of the methods approaching direct non-linear inversion. Suppose that we have defined a function measuring the fit of model field to the data. Naturally we wish to minimize this measure, the *objective function* S. Mathematically we are faced with solving the problem

$$S = f[\mathbf{F}_h, \mathbf{F}_t(\mathbf{p})] = \min! \qquad (4.1),$$

where $\mathbf{F}_h$ = vector of data values, $\mathbf{F}_t$ = the theoretically calculated field of the model and $\mathbf{p}$ is the vector of (non-linear) model parameters. S defines mathematically the "distance" between the vector functions $\mathbf{F}_h$ and $\mathbf{F}_t$ or the *norm*:

$$S = \|\mathbf{F}_h - \mathbf{F}_t\| = \|\mathbf{e}\| \qquad (4.2)$$

The vector functions $\mathbf{F}$ may be continuous or discrete and various norms define different function spaces.

To find the minimum, S is differentiated with respect to each of the parameters and each resulting equation is set equal to 0.

$$\partial S/\partial p_j = \partial f/\partial p_j = \partial \mathbf{F}_t/\partial p_j = 0 \qquad j = 1, 2, ..., M \qquad (4.3)$$

This is a set of M simultaneous non-linear equations. The methods used to find the extremum value of the objective function are called *methods of non-linear optimization*. (Note that most of the literature on non-linear optimization method is written for the purpose of financial problems or optimal control of industrial processes. In these, more often than not, the maximum of the objective function is wanted. Maximizing problems are, however, easy to transfer into minimum seeking ones. With a constant A chosen to be large enough, the problem S = min! will be equivalent to (A- S) = max!.)

Non-linear optimization can be described as a systematic mapping of the values of the components of $\mathbf{p}$ so that the extremum of S finally is found. Although a wealth of different methods exist, the number of approaches useful in practice is limited. In fact, quite a few of the most popular methods result in mathematically very similar or even exactly the same solutions as when the inversion problem is linearized and its solution is stabilized.

4.1.1 Objective Functions and Norms

The Euclidean length is the length measure which is most familiar to us from everyday life. This measure can conveniently be used to describe the length of vectors, too. There exist, however, many other possible measures of vector length. Mathematically the *norm* of a vector, e, can be defined and is denoted by $\|e\|$. The definition of a norm allows us to use it also as a measure of length of a continuous function. Such a measure is convenient to determine the "distance" between measured and modelled data in geophysical inversion. The most commonly employed norms are based on the sum of the nth power of the vector or function elements. The L_n norm for a discrete function in a linear vector space L_n is defined as:

$$\| e \|_n = \left[\sum_i^N |e_i|^n \right]^{1/n} ,$$

(4.4)

and correspondingly for a continuous function in function space L^p:

$$\| f \|_p = \left[\int_a^b |f(x)|^p \, dx \right]^{1/p} .$$

(4.5)

4.1.1.1 Exact Fit

If a solution to the inverse problem is attempted as an exact fit of the data and model fields, then the L_1 norm is preferable:

$$\| e \|_1 = \left[\sum_i^N |e_i| \right] .$$

(4.6)

This norm has found applications e.g. in experiments with analytical inversion procedures.

4.1.1.2 Least Squares

If the vector e is defined as $e = F_h - F_t$ the difference between measured data F_h and the model field F_t, then the L_2 norm:

$$S = \| e \|_2 = \left[\sum_i^N |e_i|^2 \right]^{1/2} ,$$

(4.7)

will be the natural objective function to be used in a least-squares solution of geophysical inversion. The choice implies that the data obeys Gaussian statistics, i.e. the inaccuracy of the data can be described by a normal distribution. Often in practice S^2 is used, but it is preferable instead to divide S by $\sqrt{N}$ to obtain an objective function which is comparable to standard Gaussian statistics. The popularity of this objective function or the choice of the least-squares method obviously has to do with the fact that the theory is simple and well known since Gauss devised it. Another advantage of the LSQ norm in geophysical inversion problems is that the objective function approaches linear dependence on the parameters close to the optimum (e.g. in the examples of S maps to be discussed later, the behaviour, in fact, is quadratic, since the maps were drawn for the contours of $S^2 = |e_i|^2$).

4.1.1.3 Robust Methods. The Minimax Solution.

Successively higher norms (n larger) give the largest elements of e successively larger weight. In the limit n -> ∞, the length of e will be the element with the largest absolute value:

$$\|e\|_\infty = \max_i |e_i| \qquad\qquad (4.8)$$

When this is used in inversion as the objective function, the smallest value of which is attempted, one speaks of the *minimax* solution. The mathematical theory of the minimax fit has not been well developed and many disadvantages have been attached to it. The approximation of functions with the Chebysheff polynomials is based on the minimax norm. Recently, strong emphasis has been given to robust methods, i.e. methods which can reasonably well handle data outliers (see also Fig. 4.1). The minimax solution belongs to the category of robust methods. It differs from the higher order norms, like the LSQ solution in that the difference vector e often has a successive alternating component (the neighbouring components of e are of opposite sign).

4.1.1.4 Weighted Objective Functions

There are often reasons why the error vector e alone is not sufficient to define a suitable objective function. The data may have different accuracies, and one may wish to determine the model parameters which vary smoothly from component to component etc. For such situations, the objective function can be defined as:

$$S_{weight} = e^T \underline{W} e, \qquad\qquad (4.9)$$

where the weighting matrix $\underline{W}$ differs from case to case. Menke (1989) has given a good description of the various weighting functions and their effects in the context of linear parameter inversion. He points out that a weighted objective function is no longer a proper norm, since the matrix may have negative components so that the positivity condition of a norm may be violated. Menke has defined a priori norms and flatness matrices:

$$\underline{W}_{flat} = \begin{bmatrix} -1 & 1 & & & & & \\ & -1 & 1 & & & & \\ & & & \cdot & \cdot & & \\ & & & & \cdot & \cdot & \\ & & & & & \cdot & \cdot \\ & & & & & -1 & 1 \end{bmatrix} \qquad (4.10)$$

that would lead to the overall flat objective function:

$$S = [\underline{W}_{flat} \ e]^T \ [\underline{W}_{flat} \ e] = e^T \ \underline{W} \ e. \qquad (4.11)$$

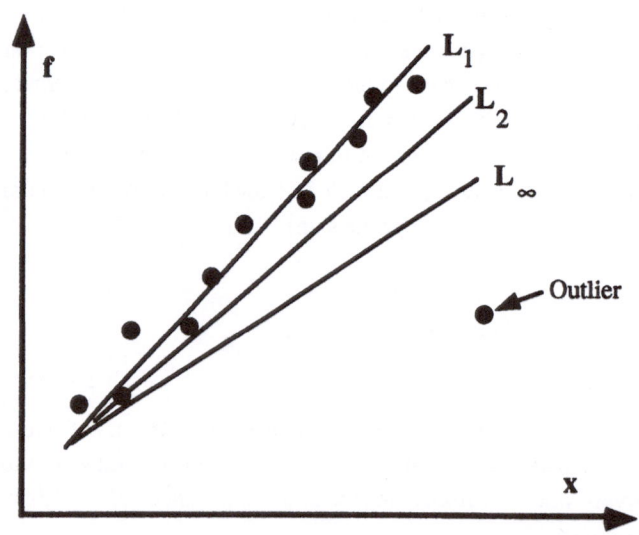

Fig. 4.1. Fitting a line to a data set under three norms L_1, L_2 and L_∞. The L_1 norm gives the least weight to the one outlier point. (After Menke 1989)

4.1.2 Properties of Objective Functions

It is not very common to do systematic research into the behaviour of the objective functions for a certain geophysical inversion model and/or method. There are, however, some general properties, which are useful to study to some extent or aim at, if one has to design a program or system of programs for repeated and/or systematic interpretation practice. Most non-linear optimization methods require that the objective function is unimodal, convex, easy and quick to calculate and that the model parameters are as independent of each other as possible. One of the toughest questions in optimization theory is how to secure a search strategy, which makes sure that, the smallest of several possible minima is obtained. Any minimum point, where the objective function has a smaller value than in its neighbourhood, is called a *local minimum*. The smallest S-value of these (in a complicated case rather numerous) minima, is called the *global minimum*. In Fig. 4.2, the minimum marked ω is a global one, those marked V and W not to speak of those unmarked, are local minima. The S-map of Fig. 4.2 is for a rather simple interpretation problem: to determine the bedrock topography from a gravity profile.

When the data contained no error, the minima ω, V and W were distinct, but still difficult to differentiate from each other reliably in a completely automatic process. The models of a topography of the bedrock /or mantle of these three minima differed only in minor details. Al-Chalabi (1970) has also shown by means of examples, that if a data profile is too short with respect to the relevant parts of the subsurface model, multimodality of the objective function becomes a serious obstacle.

A function is said to be *unimodal*, if there is some path from every point to the optimum point along which the function continually increases or decreases (Cooper and Steinberg 1970). For

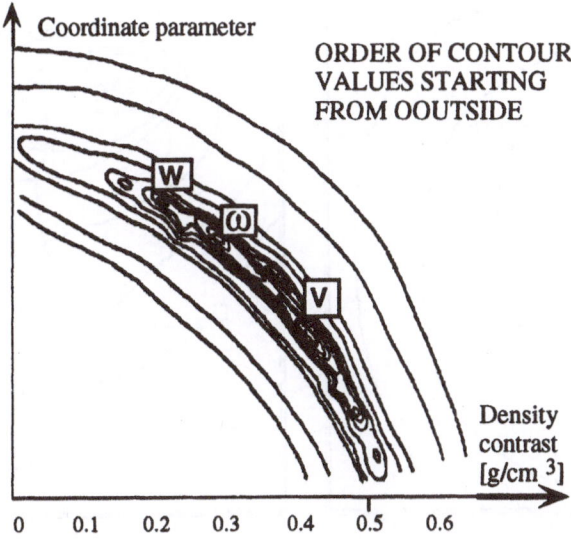

Fig. 4.2. A cross-section of an objective function in gravimetric interpretation, when the bedrock topography is described by a polygonal boundary. The example is a theoretical construction, the addition of 1 mgal rms noise has produced a multitude of local minima. (Al-Chalabi 1970)

unimodal objective functions almost any optimization strategy works and the global (in this case only) minimum is found. Unimodality is thus a highly desirable property with deep connections to the fundamental question of optimization theory, uniqueness = how to reliably find the global minimum. Figure 4.3 shows a simplified example (modified from Draper and Smith (1966)) which depicts two maps of equivalue curves of the objective function S for two arbitrary parameters. Suppose that optimization strategy is defined so that the search proceeds always in the direction of decreasing S. Then, if S is unimodal, there is a unique minimum point in the parameter space (Fig. 4.3A) and the search will find this minimum point irrespective of where in parameter space the search starts. If the objective function is multimodal, there exist several (local) minima and the minimum found will depend critically on the starting point in the parameter space. In Fig. 4.3B, showing a bimodal case, the starting point A leads to the lower (global) minimum, if the same strategy as described is followed. From starting point B the strategy brings the search to the second, less favourable local minimum. Because of the saddle point structure of the S-map, starting point C can lead to either of the minima, depending on how well the starting direction is defined.

Convexity of the objective function is also a desirable feature, which is closely connected to the unimodality. If the contours of S start to become concave, there soon develops multimodality as is easily understood by looking at the example in Fig. 4.4. The contours are obtained by mapping two different S-functions for the magnetic 2-D plate model around the true values of the plate parameters dip (horizontal axis) and thickness/depth of upper surface (vertical axes). In the upper (nicely convex) part, the square root of the sum of squares was used and the maximum error in the lower (slightly concave) part of the figure.

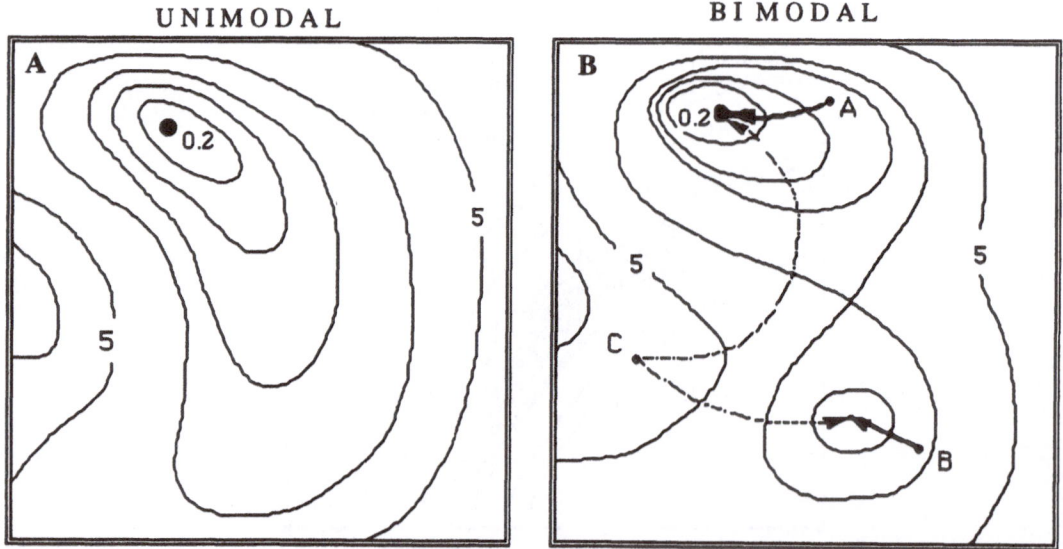

Fig. 4.3. Maps of objective functions. **A.** A unimodal function. **B.** A bimodal function

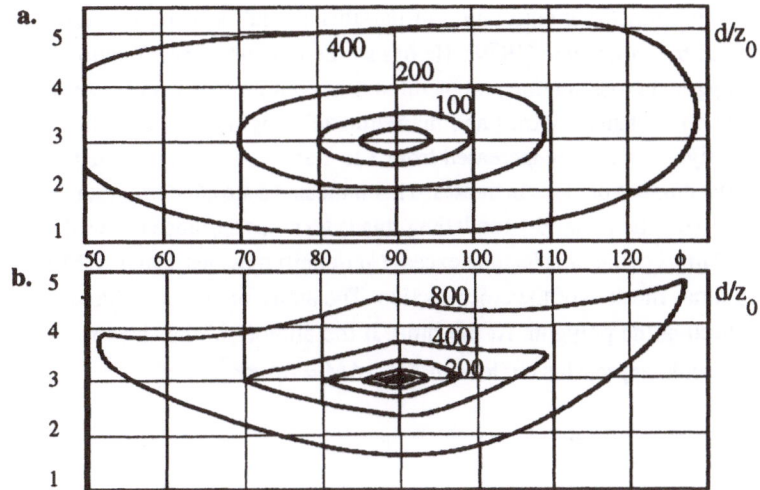

Fig. 4.4. Convex-concave behaviour of objective functions. **a.** A well-defined, convex function. **b.** A function with tendency towards concavity. (Hjelt 1973)

Correlation between the parameters occurs, when their effects on the theoretical anomaly profile are similar. Strong correlation rotates the contours of the objective function (Fig. 4.5) and the search for minimum in the direction of the parameters becomes tedious. One solution is to use intermediate variables and perform the search in the directions defined by them. In Fig. 4.5, case 3 this would be true for the variables q_1 and q_2 defined by:

$$q_1 = p_1 \cos \Phi + p_2 \sin \Phi$$
$$q_2 = - p_1 \sin \Phi + p_2 \cos \Phi$$

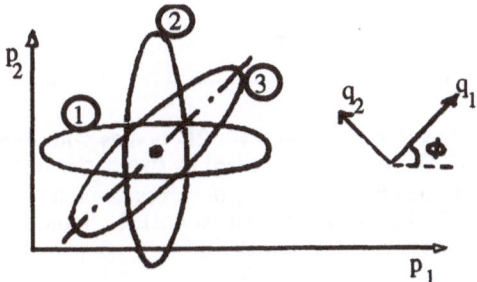

Fig. 4.5. Correlation between parameters as indicated by contours of the objective function. **1.** and **2.** No correlation; **3.** strong correlation. A search in the direction of the intermediate variables q_i would be effective. (After Draper and Smith 1966)

An elegant exposition on strong correlation of parameters and in fact on nonuniqueness of gravity inversion, is given by Al-Chalabi (1970). He has studied the objective function S of a horizontal cylinder with polygonal cross-section and pictured S as contour maps. Two parameters have been varied in the maps, the density contrast and the radius of a regular polygon. In Fig. 4.6 the number of sides of the polygon has been increased from 3 to 4 and 6. The S-contours become more and more elongated as the number of sides increases. This indicates a strong correlation between the parameters. For a person versatile in gravity modelling this is no surprise, since the well determined parameter of a horizontal mass cylinder is its total excess (or deficit) mass per length $\Delta M/L$, which is proportional to density times the area of the cross-section. The areas, on the other hand, are proportional to the radius encircling the polygon. At the limit, if the number of the sides were still to be increased, the boundary would approach a circle with $\Delta G \sim \Delta M = \Delta \rho\, \pi R^2$.

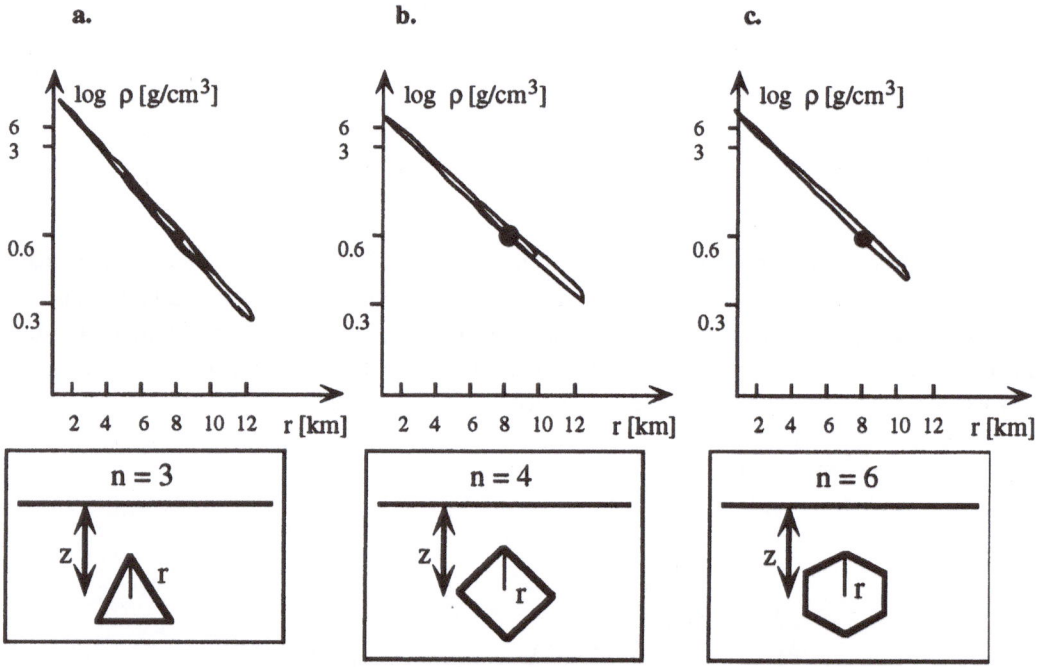

Fig. 4.6. Contours of the objective function S for 2D bodies with regular cross-section in non-linear gravity inversion. The model is descibed by two parameters, the density contrast (0.6 g/cm^3) and the radius of the circle encircling the cross-section (r = 8 km). The centre of the model is located at z = 16 km. **a.** Triangular (n = 3), **b.** Square (n=4) and **c.** Regular sextuple (n=6) cross-section. (After Al-Chalabi 1970)

Very few mappings of the objective functions for geophysical basic model bodies have been published. For the magnetized plate, Hjelt (1973) has presented an analysis, which includes objective function maps of ten parameter pairs. These maps are reproduced in Fig. 4.7. The study of the behaviour of the objective function with the parameters is seldom so much of a problem of calculation, but very much a problem of visualization. Since the simultaneous effects of more than three parameters are impossible to show in one picture, Hjelt (1973) decided to present S-contour maps of combinations of parameter pairs in a study in connection with the design of a computerized system for magnetic inversion using the 2-D plate model.

Originally, the maps were computed to study the proper amount of change for each parameter in a non-linear optimization algorithm, where optimization was performed in the original parameter space, one parameter at a time. The maps, however, give insight also into many other properties of the model parameters and their behaviour. For example the maps indicate, that thickness d should be mapped with a step change which is twice (Fig.4.7 j) and depth extent h with a step about six times (Fig. 4.7 c) the amount of horizontal position x_0. (Note that the parameters x_0 and z_0 are given as x and z in the figures and in the figure legend for reasons of simplicity.) When the contours of the S-value are inclined with respect to the parameter axes, corresponding parameters are correlated as explained. This is true for the magnetic plate model for the parameter combinations x_0 - ϕ (position and dip, Fig. 4.7 f), d - z_0 (thickness and depth, Fig. 4.7 i) and d - h (thickness and depth extent, Fig. 4.7 d). These parameters should preferably be changed together (e.g. as a linear combination as explained in connection with Fig. 4.5) rather than separately.

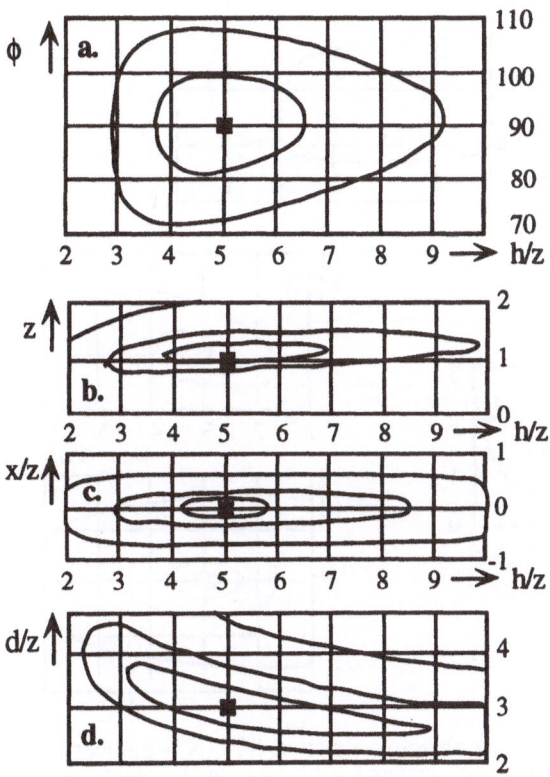

(Fig. 4.7 continues)

(Fig. 4.7 continued)

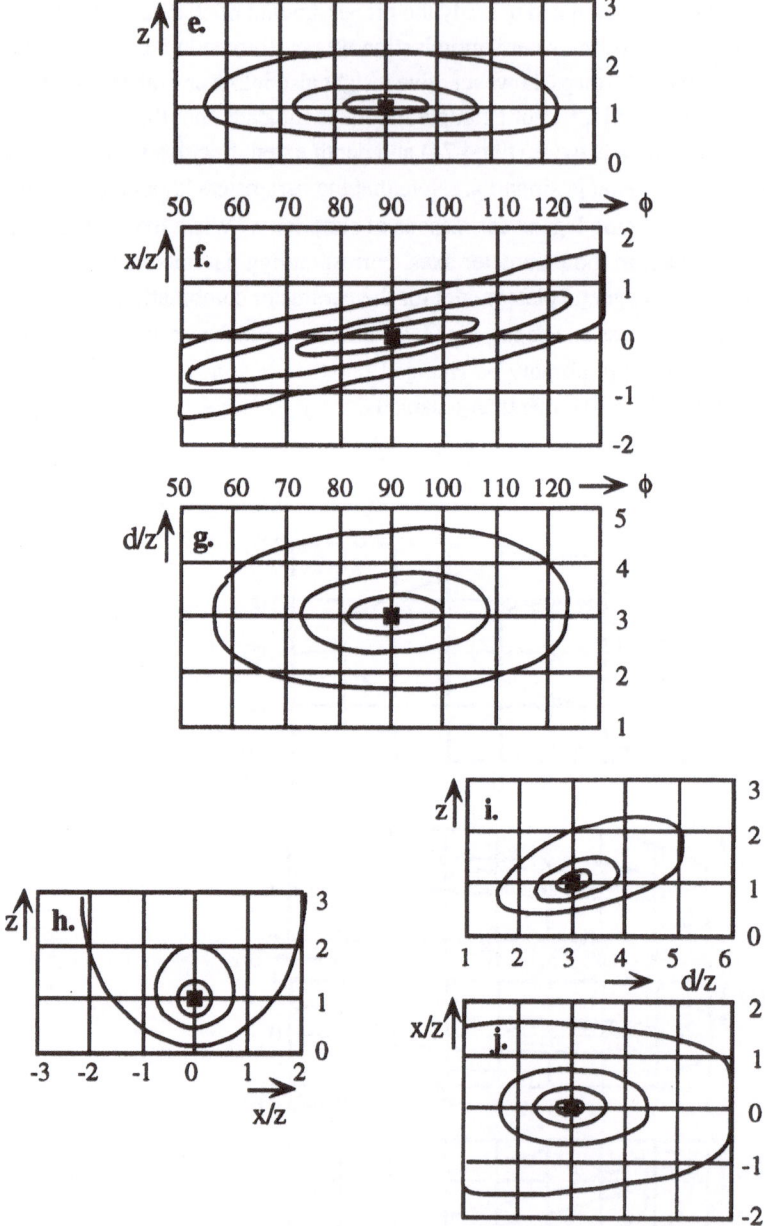

Fig. 4.7. Maps of sum of squares objective function S/N (N = number of data points) of a 2-D magnetized thick plate. A pair of parameters has been mapped around the true value of these plate parameters, indicated by the **cross** in the middle of each map. The parameters are: **d** = thickness, **h** = depth extent, **x** = horizontal location, **z** = depth to the upper surface, φ = dip, **k** = susceptibility.The S/N curves for 0.25, 1, 4 (and 10) are shown. **a.** φ - h/z; **b.** z - h/z; **c.** x/z - h/z; **d.** d/z - h/z; **e.** z - φ; **f.** x/z - φ; **g.** d/z - φ; **h.** z - x/z; **i.** z - d/z; **j.** x/z - c/z (Hjelt 1973)

Some parameters should be changed logarithmically (in proportion to their instantaneous value), since the S-contours are elongated towards higher numerical values of the parameter. The depth z_0 and the depth extent h are such parameters (see. Fig. 4.7 b, e, h). Strong non-linearity (in addition to correlation) is evident for the parameter combination d - h (thickness and depth extent, Fig. 4.7 d). Some degree of non-linearity appears for the parameter combination ϕ - h (dip and depth extent, Fig. 4.7 a) and z - h (depth and depth extent, Fig. 4.7 b). An almost ideal second-degree behaviour of the S-function holds true for the combination d - ϕ (thickness and dip, Fig. 4.7 g).

4.1.3 Constrained Optimization

Optimization methods need to include a mechanism whereby any parameter can be restricted to a certain range of possible values, e.g. electrical conductivities and seismic velocities are always positive, whereas the logarithm of the conductivity, density and even magnetic susceptibility can be negative also. If the depth to the top of a body is a model parameter it by necessity also has to be > 0, otherwise a nonsense result of a body protruding from the Earth's surface would result.

There is no general theory for applying constraints to parameters in non-linear optimization. The approaches range from the method of Lagrangian multipliers to the use of "penalty" functions. Constraints can be of two types, equality and inequality ones. In geophysical inversion problems the latter type dominates. There does not seem to be any widely accepted choice of a best method for inclusion of constraints. The earliest approaches were simply to apply barrier terms to restrict the domain of variation of the parameters. Modern methods tend to treat the constraints as equations and to incorporate them into the main minimization problem.

4.1.3.1 Penalty Functions

The basic idea behind using the penalty function approach is to attempt to transform the non-linear constrained optimization problem into a single or a sequence of unconstrained problems. A penalty function can be (Fletcher 1982): *sequential,* (= the solution of the constrained component(s) of **p** is located as a limit of a sequence of unconstrained penalty function minimizations) or *exact* (= the solution of the constrained component(s) of **p** is the minimizer of a single penalty function). There is no general rule for the efficiency to distinguish between these two variants, although many aspects seems to favour the use of exact penalty functions. For the simple penalty function approach the objective function is mathematically continued outside the permissible region of the parameter by using a multiplying function, e.g. of the form:

$$S_{ext} = S\,[1+ w\,(p_j - p_{j,\,bound})] \quad \text{with} \quad w = |p_j - p_{j,\,bound}|^n \;\text{or}\; \exp|p_j - p_{j,\,bound}|. \quad (4.12)$$

The best result for the optimization technique is obtained when at least the first derivative of S_{ext} is continuous across the relevant parameter boundary. Experiments made during the construction of a magnetic interpretation program package (Hjelt 1973) indicated that a power function where n=2 was better than an exponential penalty function, at least when used in combination with second degree methods of search for the optimal parameter values.

4.1.3.2 Lagrangian Multipliers

Fletcher (1982) and Conn (1982) have discussed several other variants of constraining the parameters. In the sequential penalty function or the augmented Lagrangian approach of Fletcher S_{ext} is written as a combination of first and second order terms of w in Eq. (4.12) with $n = 1$:

$$S_{ext} = S - \lambda^T w(p) + (\sigma/2) \, w^T(p) \, w(p). \tag{4.13}$$

The Lagrange-Newton technique is often used in Quadratic Programming optimization. The choice of the penalty parameters is critical for the success of any penalty function method. There is often a strong conflict between the aim of minimizing the original objective function and of satisfying the parameter constraints. Conn (1982) suggests that parallel processing computer techniques may alter totally the choice of suitable penalty functions. The virtues of hybrid methods, which means using different approaches to the constraint problem during various stages of the minimization procedure, were also pointed out in the discussion section of constrained optimization in Powell (1982).

To illustrate constraint optimization the following simple example is repeated from Cooper and Steinberg (1970).

Problem: Min $S(p) = 2p_1^2 + 3p_2^2$

subject to the constraint: $p_1 + 3 \cdot p_2 = 10$ $\tag{4.14}$

Solve first from the constraint equation:

$p_1 = 10 - 3 \cdot p_2$

and substitute into the objective function S. Then one has a single-parameter minimization problem

$$\text{Min } S(p) = \text{Min } \{2(10 - 3p_2)^2 + 3p_2^2\} = \text{Min } [h(p_2)] . \tag{4.15}$$

If a solution cannot be easily found, then e.g. method of Lagrangian functions needs to be applied. Define:

$$F(p, \lambda) = S(p) + \lambda[b - g(p)]. \tag{4.16}$$

The solution is sought by solving simultaneously:

$$\frac{\partial F}{\partial p} = 0$$

$$\frac{\partial F}{\partial \lambda} = 0 \tag{4.17}$$

In a recent paper Whittall (1986) has discussed the inversion of magnetotelluric (MT) data using constraints for the conductivity. His algorithm is based on linear programming to solve the non-linear Riccati differential equation governing the behaviour of the complex reflectivity of the plane wave electromagnetic fields of magnetotellurics. The model consists of a one-dimensional, layered Earth, the conductivities and depths of the layer boundaries being the model parameters. Measurements are assumed to be taken at eight discrete frequencies.

The result of iterative inversion of the data is presented in Fig. 4.8. The expositions are slightly modified from the original in the paper by Whittall. In the upper Fig. the curve UC denotes the un-constrained model achieved after three iterations starting from a homogeneous Earth model (with the conductivity the same as for the uppermost layer, 0.002 S/m). The result does not depict the detailed structure of the true model.

When the conductivities of the layer with the highest conductivity, of the resistive layer around $z = 300$ and of the bottom layer are constrained to values 5 to 10% of their true values (the band is indicated by the thicker bars on the layer model conductivity in Fig. 4.8 a, b), the inversion results in a rather good description of the structure of the layer model. The conductivities of the result do, however, vary considerably with respect to their true values.

4.1.4 Stopping Criteria in Iterative Optimization

Most of the non-linear optimization methods are iterative in one way or another. The minimum of the objective function is approached through successive steps of improvement. Computer reali-zations of various algorithms require definite criteria for when the fit is acceptable and the iteration process can be stopped. This is especially essential in more complicated cases, when the number of model parameters is high. It would be rather natural to assume that an acceptable limit of fit is achieved when the objective function is statistically comparable with the average error of the data (or with average variance when a quadratic norm is used). Many other criteria have been proposed and used in practical computer programs.

We distinguish between four major types of *stopping criteria:*

$$
\begin{aligned}
&1.\ S < S_0. \\
&2.\ \Delta S < e_S \\
&3.\ \Delta p_i = e_{pi} \qquad \text{for all } i = 1, 2, ..., M. \\
&4.\ \#\ \text{iter} \ge \text{MAXIT}
\end{aligned} \qquad (4.18)
$$

The first criterion has a great statistical appeal, but the proper determination of the value of S_0 is far from being easy. It is partly related to the question of finding a global minimum instead of a local one. Selecting a too stringent criterion forces an optimization program to try to improve the search in the surrounding of a local minimum, where no great improvements are possible. Voluminous experience with magnetic profile interpretation (Hjelt 1973, 1975) showed that for this type of stopping criteria, the running times often increased considerably while the search was concentrated on small improvements of less important parameters of the model. This was especially so for long profiles, with many different model bodies and a great number of model parameters.

a.

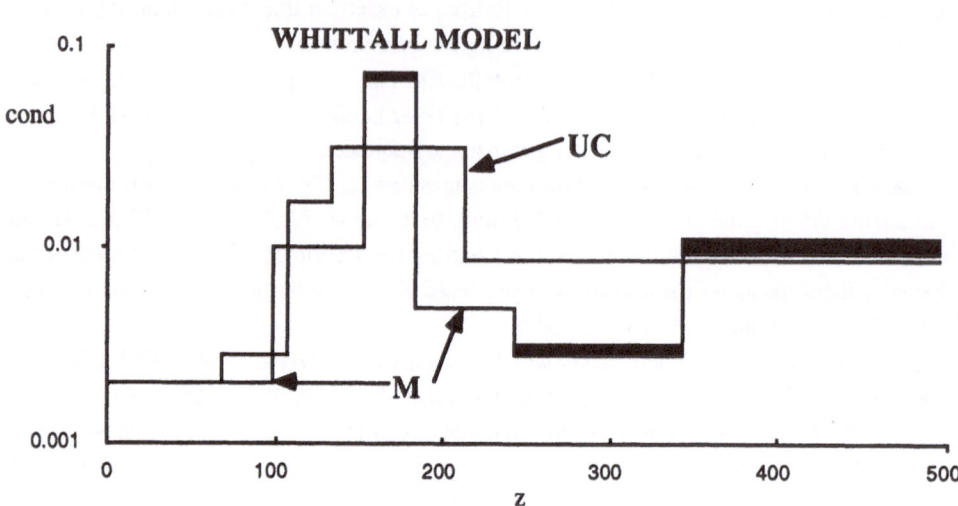

Fig. 4.8 a. Unconstrained optimization of MTS data of a six-layer (theoretical) model. M = Original model; UC = optimized, unconstrained model (after Whittall 1986)

b.

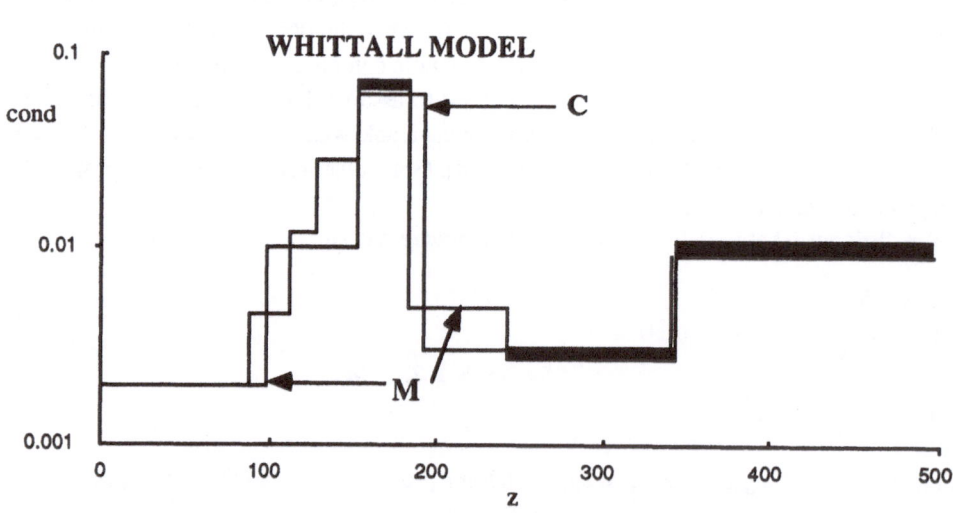

Fig. 4.8 b. Constrained optimization of MTS data of a six-layer (theoretical) model. M Original model; UC optimized, constrained (seven-layer) model (after Whittall 1986)

The second criterion can also be easily motivated statistically. Experience indicates that this criterion tends to stop the iteration too early. The objective function often behaves during an iteration process, so that its decrease slows down or almost ceases temporarily. This can be due to a search around a local minimum, which after a few iterations will be abandoned and a new greater pace of decrease starts again. Such a behaviour seems to be rather common for multiparameter problems and more complicated models.

The third criterion is related to the resolution of the model parameters. It has been used rather widely, and it has an appealing property that each parameter can be given a different limit value e_{pi} which is typical for the corresponding parameter. In many geophysical interpretation programs it is often desirable to fix the value of one or several parameters. The fix can be permanent, because the value of the parameter/parameters is/are known from other sources. Alternatively, a temporary fix may be wanted, for one reason or another. Such options are most easily implemented in connection with this stopping criterion.

Most modern implementations of geophysical interpretation programs are done on work stations or personal computers in order to allow for an interactive operation mode of the optimization procedures. The fourth stopping criterion is very well suited for such implementations allowing a possibility of controlling the process of the iterative inversion and often making considerable savings in computing time. An efficient algorithm to search for the optimum parameters for minimum objective function value for a few iterations, making a pause in the search and looking at the result, making corrections to the model based on the experience of the interpreting geophysicist is an ideal and cost-effective way to handle inversion in practice. This seems to be a reasonable approach to the difficult question of global versus local minimum, so crucially important also in geophysical interpretation.

4.2 ONE-DIMENSIONAL OPTIMIZATION

The basic idea in optimization methods is a systematic search of the parameter values, so that the global extremum of the objective function is found. This search involves the repeated computation of the model field (and/or its derivatives, if the chosen optimization method happens to require them for operation). The easiest and the most objective measure of the efficiency of optimization is the necessary number of function determinations to reach a given level of fit between data and model fields.

Although effective search methods operate in multidimensional space, it is instructive to first understand some basic properties of one-dimensional search procedures. Suppose, that by a procedure we have arrived at the situation, where the value of the parameter p giving the optimal (minimum) value for S is in the region $L_0 = <a, b>$. In order to find a new, smaller region of variation, L_1, two new objective function determinations are needed (Fig. 4.9).

Let $S_1 = S(p^{(1)})$ and $S_2 = S(p^{(2)})$, then obviously:

$$L_1 = <p^{(1)}, b> \qquad \text{if } S_1 > S_2$$
$$L_1 = <a, p^{(2)}> \qquad \text{if } S_1 < S_2.$$

The choice of the new parameter values $p^{(1)}$ and $p^{(2)}$ with respect to a and b, leads to various one-dimensional optimization strategies.

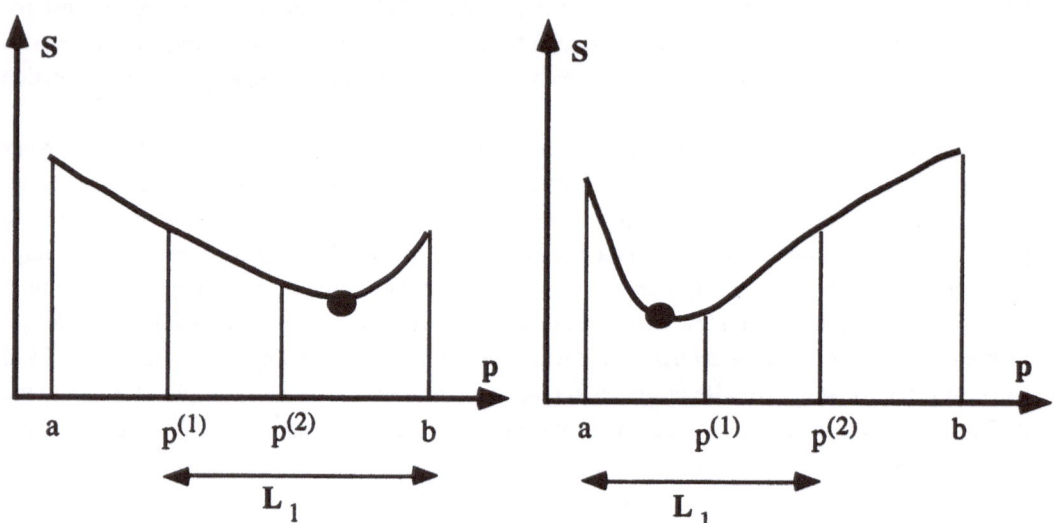

Fig.4.9. Two objective function calculations are needed to determine the new parameter region L_1.

4.2.1 Golden Cut

Let us aim at such a choice of the new parameter values that $L_{i+1}/L_i = q = \text{const}$. We then have:

$$\frac{p_i^{(2)} - a_i}{b_i - a_i} = \frac{b_i - p_i^{(1)}}{b_i - a_i} = q, \qquad (4.19)$$

which after rearranging can be written:

$$p_i^{(1)} - a_i = b_i - p_i^{(2)} = b_i - a_i - (p_i^{(2)} - a_i). \qquad (4.20)$$

By making systematically the choices:

when $S_2 > S_1$:

$$p_{i+1}^{(2)} = p_i^{(1)}, \ b_{i+1} = p_i^{(2)} \ \text{or}$$

when $S_2 < S_1$:

$$p_{i+1}^{(1)} = p_i^{(2)}, \ a_{i+1} = p_i^{(1)},$$

one obtains:

$$\frac{p_{i+1}^{(2)} - a_i}{p_i^{(2)} - a_i} = \frac{p_i^{(1)} - a_i}{p_i^{(2)} - a_i} = q. \tag{4.21}$$

By a substitution of $p_i^{(1)}$ from Eq (4.20):

$$q = \frac{b_i - a_i - (p_i^{(2)} - a_i)}{p_i^{(2)} - a_i} = (1/q) - 1. \tag{4.22a}$$

The solution of this equation of second degree for q is:

$$q = (\sqrt{5} - 1)/2. \tag{4.22b}$$

The new parameter values $p^{(1)}$ and $p^{(2)}$ have to be chosen so that the line <a, b> is divided by the classical *golden cut*. After N search rounds, we have computed 2N new objective functions S and the variation range of the parameter has been decimated to:

$$L_N = q^{N-1} L_1. \tag{4.23}$$

The golden cut is next to the optimal 1-D search procedure (Draper and Smith 1966).

4.2.2 Fibonacci Search

If we consider the decimation of a range of variation of a model parameter from the another point of view. Given L_0, the range of variation of p, one has to decimate L_0 to unit length by calculating N values of the objective function. Then it can be shown that the Fibonacci search is optimal in the sense that it allows for the greatest value of starting L_0 of all search procedures. The algorithm for this procedure is:

$$p_i^{(1)} = a_i + q_1 (b_i - a_i) \qquad \text{with } q_1 = F_{N-i-1} / F_{N-i+1}$$
$$\tag{4.24}$$
$$p_i^{(2)} = a_i + q_2 (b_i - a_i) \qquad \text{with } q_2 = F_{N-i} / F_{N-i+1}$$

Depending on the size of S at $p_i^{(1)}$ and $p_i^{(2)}$, the new range of variation will be either:

$$a_{i+1} \to a_i; \quad b_{i+1} \to p_i^{(2)} \qquad\qquad \text{if } S(p_i^{(1)}) > S(p_i^{(2)}):$$

or

$$a_{i+1} \to p_i^{(1)}; \quad b_{i+1} \to b_i \qquad\qquad \text{if } S(p_i^{(1)}) < S(p_i^{(2)}).$$

F_k are Fibonacci numbers obeying a simple addition rule:

$$F_k = F_{k-1} + F_{k-2} \qquad\qquad \text{with } F_o = F_1 = 1. \qquad\qquad (4.25)$$

The numbers are due to a 12th century monk, nicknamed Fibonacci, investigating the number of subsequent generations of rabbits. The table of the first ten Fibonacci numbers is shown in Table 4.1. When k approaches infinity, the quotient of two subsequent numbers approaches $(\sqrt{5} + 1)/2$. It is easily shown (Cooper and Steinberg 1970) that for the Fibonacci search after N rounds, the range of variation of the parameter p has decimated to:

$$L_N = L_1/F_N. \qquad\qquad (4.26)$$

Although the Fibonacci search is optimal in the sense mentioned earlier, the golden cut search needs only one objective function calculation more to reach the same decimation into unit length (starting with L_o).

TABLE 4.1.   Fibonacci numbers

k	1	2	3	4	5	6	7	8	9	10 -> ∞
F_k	1	2	3	5	8	13	21	34	55	89 -
F_k/F_{k-1} -1	-	1.000	0.5	0.667	0.6	0.625	0.6154	0.619	0.6176	0.6182 0.61803

4.2.3 Speedup by Parabolic Fit

One-dimensional search for the optimum of a single parameter can conveniently be speeded up by fitting a polynomial to the object function calculated for several parameter values. Polynomials of a degree higher than two become oscillatory rapidly, so a quadratic speedup is the only practical and recommendable procedure. As will be shown later, the procedure is easily extended to multidimensional fitting.

In quadratic or parabolic fit of one-dimensional search, the three values of the objective function S are used. A second degree polynomial $a_o p^2 + a_1 p + a_2$ is fitted to pass through the three points $(p^{(1)}, S_1)$, $(p^{(2)}, S_2)$ and $(p^{(3)}, S_3)$. The minimum of the parabola is estimated by $p^{(4)} = -a_1/(2a_o)$. If the objective function is almost quadratic this $p^{(4)}$ will be a good estimate of p_{i+1}. To be aware for situations where S varies strongly (e.g of a coarse step of search for p), one should select for $p_{(i+1)}$ that value of $(p_i^{(1)}, p_i^{(2)}, p_i^{(3)}, p_i^{(4)})$ which gives the smallest S.

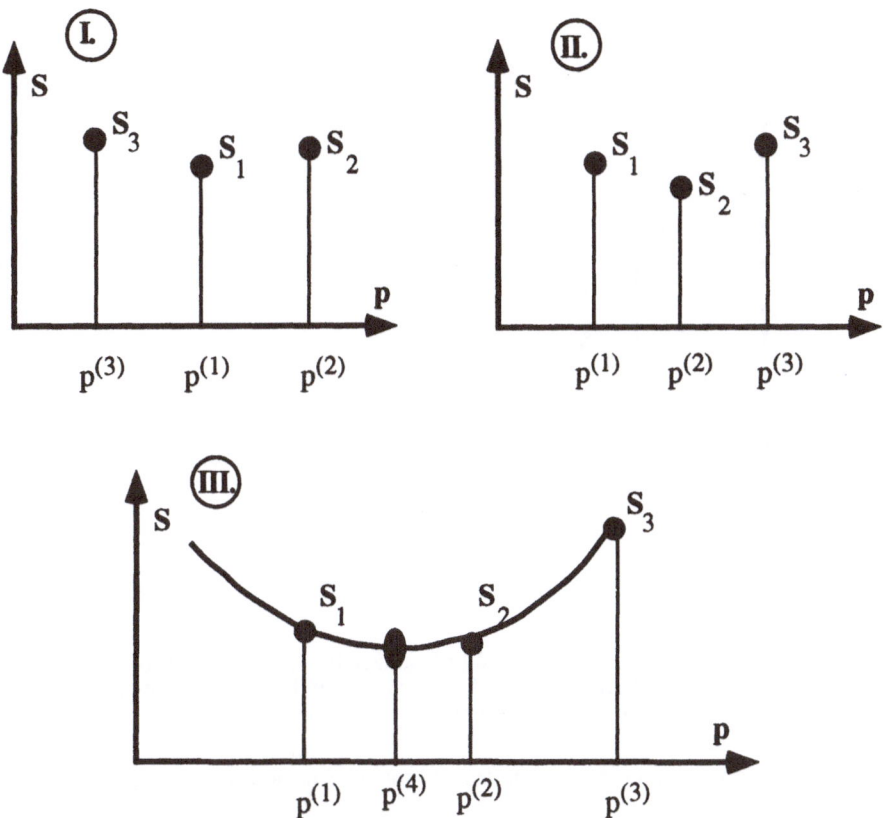

Fig. 4.10. One-dimensional optimization with quadratic fit to minimize S. For explanation, see the text

The procedure for selecting the parameter values for the S-function calculation and parabolic fit is the critical factor of quadratic speed-up of one-dimensional optimization. The simplest scheme is the following (Fig. 4.10):

$$p_i^{(2)} = p_i^{(1)} + \Delta p_i$$
$$p_i^{(3)} = p_i^{(1)} + \Delta p_i \qquad \text{if } S_2 < S_1 \qquad \text{(Step I in Fig 4.10)}$$
or $\quad p_i^{(3)} = p_i^{(1)} - \Delta p_i \qquad \text{if } S_2 > S_1. \qquad \text{(Step II in Fig 4.10)}$

If the quadratic speedup is moderately successful during one iteration, the new parameter value is at a reasonable distance from the three values used in the fit. Then it is reasonable to make the search during next iteration with a smaller step Δp_{i+1}. In the opposite case, that p_{i+1} is very far from $p_i^{(1)}$, $p_i^{(2)}$ or $p_i^{(3)}$ even an increase in the search step may become necessary. A complete iteration algorithm has to contain a selection mechanism for Δp. In practical experiments in connection with magnetic inversion with the plate model (Hjelt 1973, 1975), a constant decrease in the form $\Delta p_{i+1} = \Delta p_i / f$, with $f = 3\text{-}5$, even 10 was used. The optimal selection of Δp would be based on the distance $|p_{i+1} - p_i|$.

4.2.4 The Secant Method

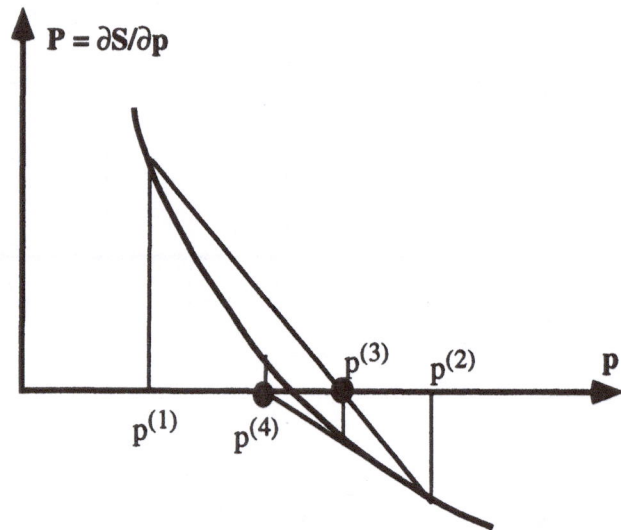

Fig. 4.11. The secant method of finding the zero of a non-linear function P is based on local linearization of P. If P is the derivative of an objective function S, the zero of P(p) is at the same time an extremum of S

Finding the extremum value of S is equivalent to solving (the non-linear) equation

$$\frac{\partial S}{\partial p} = P(p) = 0. \tag{4.27}$$

Two straightforward and much-used methods are the method of secants and Newton's method. The first is based on local linearization of P(p) and the latter on the use of the gradient of P. For the secant method (also named *regula falsi* = method of false position) the search algorithm is:

$$p_{i+1} = p_{i-1} + (p_i - p_{i-1}) \frac{P_i}{P_i + P_{i-1}} \qquad \text{with } P_i = \partial S/\partial p \ (p_i). \tag{4.28}$$

The successive application of the method is shown graphically in Fig. 4.11.

4.2.5 The Gradient Method

A great variety of optimization methods are based on the use of the gradients of the objective function. Newton's method of finding the zero of a non-linear function P does the same, but now with respect to P, which already is a derivative of S. During each iteration, the new parameter value is found by:

$$p_{i+1} = p_i - \frac{P_i}{\partial P/\partial p}\bigg|_{\text{at } p_i}. \tag{4.29}$$

A graphical exposition of the method is shown in Fig. 4.12.

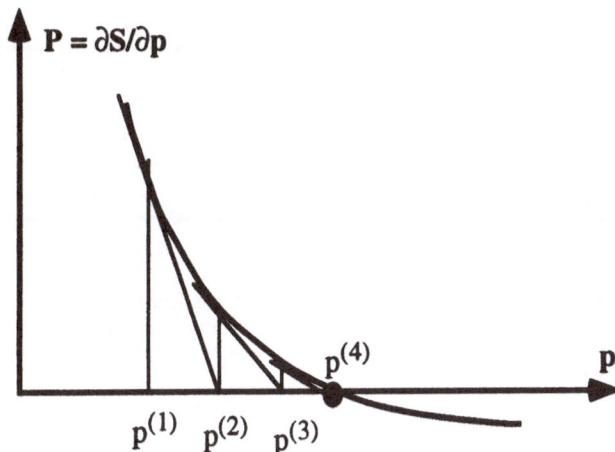

Fig. 4.12. Newton's method of finding the zero of a non-linear function P is based on the gradient of P. If P is the derivative of a function S, the zero of P(p) is at the same time an extremum of S

4.2.5.1 On the Instability of the Newton Iteration

Non-linear phenomena have come into the focus of intensive studies recently and new mathematical concepts of deterministic chaos and fractal mathematics have gained interest in various branches of natural sciences. There is an interesting connection with non-linear inversion problems and especially the linearized Newton method. In their book Becker and Dörfler (1989) describe the properties of Newton's method for finding the zero(es) of a non-linear function. Drawing on a parallel from the stability properties chaotic systems they consider each of the zeroes as an attractor and describe by means of a simple third order polynomial how different initial values lead Newton's method to converge towards different zeroes (attractors).

Following the idea and presentation of Becker and Dörfler (1989) the regions of convergence or so called basins of attraction were studied by starting the Newton iteration from successive points first along the whole region shown in Fig. 4.13. Between the "naive" or "natural" basins there is a small region where not the neighbouring zero is reached, but the zero on the opposite side (from the boundary between "naive" regions 1 and 2 to zero # 3 and and from the boundary between "naive" regions 2 and 3 to zero # 1). This behaviour can be easily studied graphically (for further details consult Becker and Dörfler 1989) and it is one form of manifestation of the numerical sensitivity of gradient methods discussed earlier.

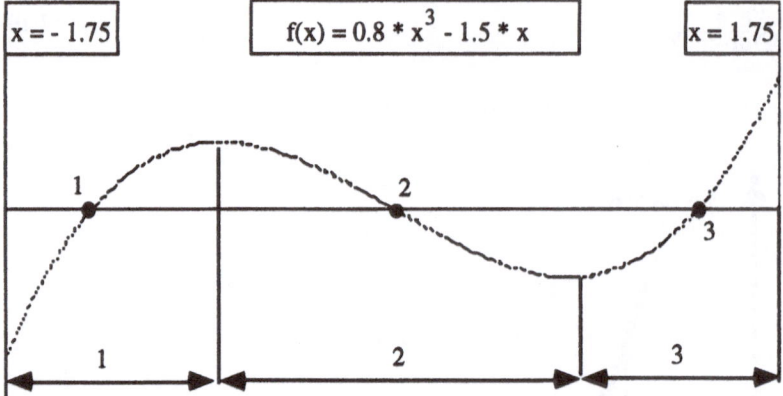

Fig. 4.13. A third degree polynomial and its three zeroes (numbered dots). The "naive" regions of attraction are indicated by numbered arrows. An initial value in any of these regions could be thought to lead the Newton iteration to the zero of the corresponding region.

The interesting fact about these intermediate boundaries is that they seem to scale in a fractal way, as is demonstrated in the series of basins in Fig. 4.14. When a smaller step between initial values of the iteration is used around either intermediate boundary, the pattern repeats itself: not exactly but to a great degree. Decreasing the step of initial values further, the repetition continues, until numerical accuracy of the computer creates additional difficulties. For multidimensional functions the basins of attractors create fractal patterns which have become so fashionable in many popular books.

4.3 MULTIDIMENSIONAL SEARCH

The main principles of the parameter search presented for one dimension remain the same, when the number of model parameters is increasing. Multidimensional search has, however, its own complications and pecularities. Unimodality becomes a much rarer property, it is also more difficult to locate. The objective function is more cumbersome to map and to visualize.

Quite irrespective of the model and the form of the objective function map, the search becomes inefficient with the increase in the number of parameters. This general property has been coined "vastness of the parameters space" (Wilde 1964). It can be described as follows. Suppose the region containing the parameter optimum is known to be $<p^{(1)}, p^{(2)}>$ (Fig. 4.15 A) and we wish to decimate it to one tenth. For his new, smaller region $<p^{(3)}, p^{(4)}>$ then $\dfrac{<p^{(3)}, p^{(4)}>}{<p^{(1)}, p^{(2)}>} = c = 1/10.$

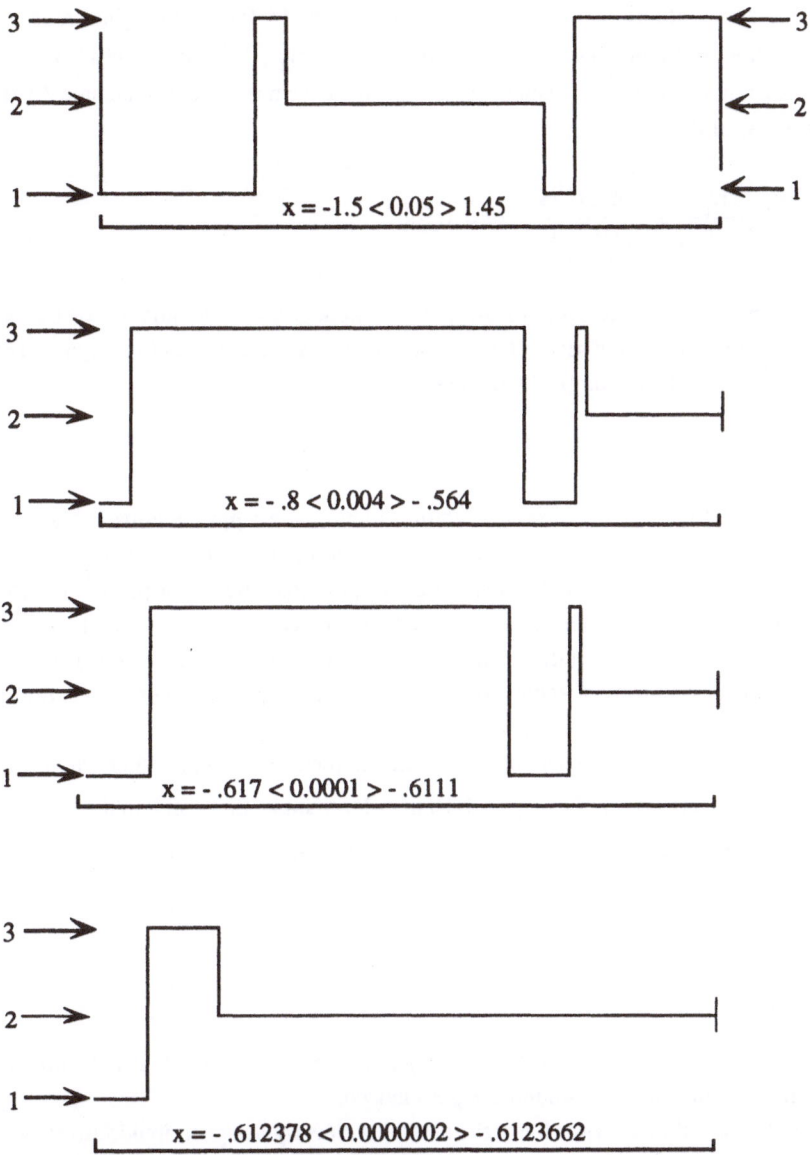

Fig. 4.14. Basins of attraction of a third degree polynomial when its zeroes are determined by the Newton iteration

Picture the same situation for two parameters (Fig. 4. 15 B). The original area where the parameters have their optimal values, is $A_0 = <p_1^{(1)}, p_1^{(2)}> x <p_2^{(1)}, p_2^{(2)}>$. If we wish to decimate the variation of both parameters with the same amount, $c= 1/10$, we need to locate an area A_1 (the black area in Fig. 4.15 B) of the size:

$$A_1/A_0 = \frac{<p_1^{(3)}, p_1^{(4)}>}{<p_1^{(1)}, p_1^{(2)}>} x \frac{<p_2^{(3)}, p_2^{(4)}>}{<p_2^{(1)}, p_2^{(2)}>} = c^2 = 1/100$$

If the number of parameters increases to M, then the homogeneous decimation of the parameter ranges with a factor c requires the search for a volume in the M-dimensional hyperspace. The size of this hypervolume divided by the original volume is:

$$V_M/V_o = c_M = c^M .$$

Already for M = 6, with c = 1/10, our search requires aiming at a 1 ppm volume in the hyperspace. At the same time the variation range (the accuracy) of each parameter has become only 10% of the original. This often is not a reasonable resolution in practical inversion problems. The multi-dimensional search definitely soon becomes the classical "search for the needle in the haystack". R. Bellman has called this the "curse of dimensionality" since according to him "... the difficulty of a n-dimensional optimization problem seems to increase in some hyperexponential fashion with the number of dimensions." (Cooper and Steinberg 1970).

The inefficiency of multidimensional search can be pictured also the other way round. If we define our aim of search to be $V_1/V_0 = c_M = 1/10$ in order to make the search in the M-dimensional hyperspace more effective. Then, again assuming that the decimation factor is the same for each of the parameters:

$$c = \frac{<p_i^{(3)}, p_i^{(4)}>}{<p_i^{(1)}, p_i^{(2)}>} = c_M^{1/M} .$$

Taking a few numerical examples, M = 3, c = 0.45 and for M = 50, c = 0.93. The decimation of the parameters, c, thus the increase in resolution, is poor indeed.

In view of the preceding analysis it is quite clear that cleverly designed multidimensional search methods will be extremely valuable. Most of the practical algorithms perform better when the objective function is strongly unimodal or exhibits linear behaviour near the optimum. In order to approach unimodality better it is often suggested scaling the variables so, that the steps of change of the parameters are of equal size. Scaling can on the other hand also be used to emphasize some of the parameters. Automatic scaling may lead to conflicts between these requirements (Powell 1981).

The problem of assuring the global extremum instead of a local extremum cannot be solved by any elegant mathematics (Powell 1981). Instead probabilistic or heuristic approaches are required. Modifying line search into curvilinear search and trust region algorithms are among proposed solutions to the global extremum problem.

There are a great variety of ways to classify multidimensional search methods. The algorithms to be described below can be classified as:

direct (univariate) search methods
>   sequential search
>   hyperparabolic fit (s.s. with quadratic improvement)
>   pattern search

multivariate search methods
>   simplex
>   conjugate directions
>   steepest descent methods

Only the main features of the methods are given next. For further information about convergence and other mathematical proofs, the reader is referred to many of the excellent books on the subject of optimization.

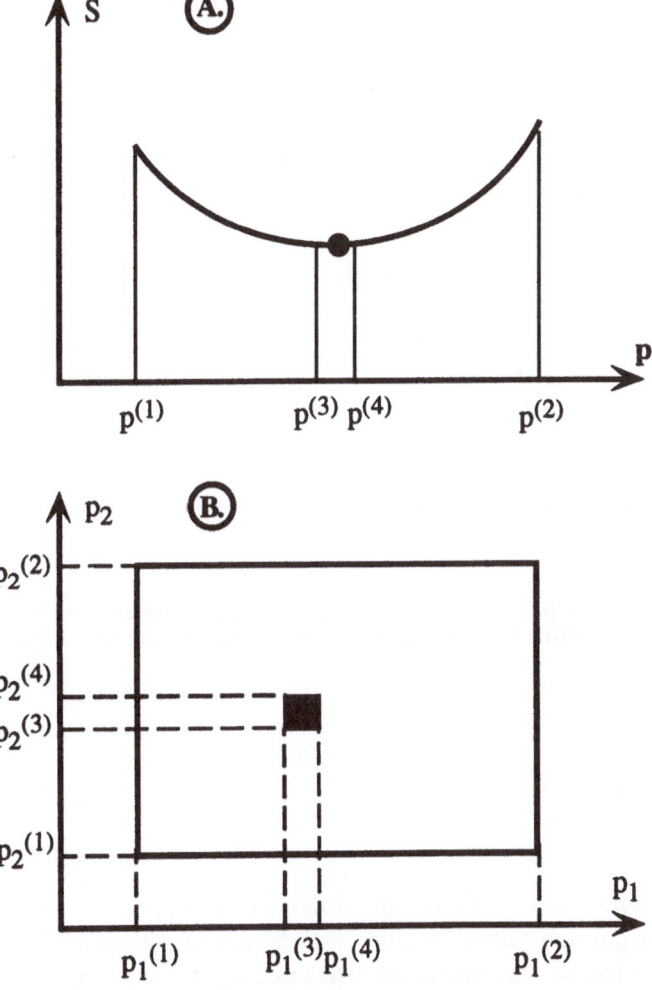

Fig 4.15. The "vastness" of parameter space. **A.** The space of one parameter. The region around the **black dot** contains the optimal value of p. **B.** The space of two parameters. The black region contains the optimal value of p_1 and p_2

4.3.1 Direct Search

4.3.1.1 Sequential Search (Search by Parameters)

The simplest and most straightforward method to search for the minimum of the object function is to search for the parameters one after another (Fig. 4.16). After having made a search once for all parameters, the procedure can be repeated using modified search steps in comparison to the previous iteration. The search for any of the parameters may be effected either using predetermined steps or interpolation techniques. Quadratic or higher degree speed-up has been described in Section 4.2.3. One of the major disadvantages of this technique is that it is difficult to follow a strongly curved objective function.

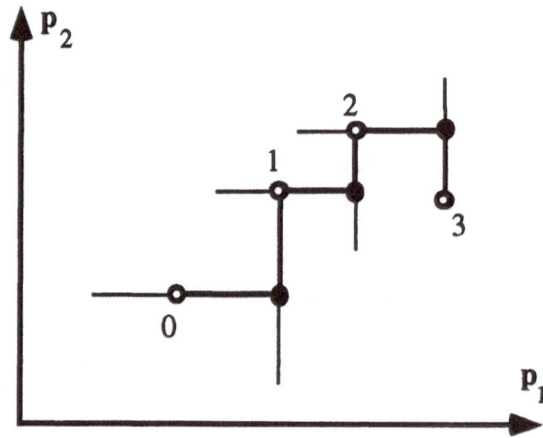

Fig. 4.16. Direct search in the space of two parameters. **Open circle:** starting point of iteration; **numbers:** iteration #; **thin line:** direction of unsuccessful search; **thick line:** direction of successful search

4.3.1.2 Hyperparabolic Fit

As in the one-dimensional case the search for the optimal value of the parameters can be speeded up by higher-degree approximation of the object function S. Lakanen (1975) has extended the parabolic fit system into several dimensions. The basic equation for a hyperparabola in n+1 dimensions can be written as:

$$S = a_1 p_1^2 + a_2 p_2^2 + ... + a_n p_n^2 + a_{n+1} p_1 p_2 + ... + a_j p_{n-1} p_n + a_{j+1} p_1 + ...$$
$$.... + a_k p_n + a_{k+1}. \qquad (4.30)$$

with $j = n(n+1)/2$ and $k = n(n+3)/2$. In order to determine the coefficients one needs $N = k+1$ equations, i.e. N values of the function S.

To accomplish this, each of the parameters is given values p_{io}, $p_{io} + d_i$, $p_{io} - d_i$, where d_i is the (predetermined) step of change for each of the parameters. Let:

$$S_1 = S(p_o)$$
$$S_2 = S(p_o + d_1)$$
$$S_3 = S(p_o - d_1)$$
$$S_2 = S(p_o + d_2)$$
$$S_3 = S(p_o - d_2)$$

.....

$$S_{2n+1} = S(p_o - d_n)$$
$$S_{2n+2} = S(p_o + d_1 + d_2)$$
$$S_{2n+3} = S(p_o - d_1 + d_2)$$

....

$$S_N = S(p_o - d_{n-1} - d_n) .$$

(4.31)

The basic equation will be in matrix form:

$$\mathbf{S} = \mathbf{H} \, \mathbf{A}^T,$$

(4.32)

where $\mathbf{A}$ is the vector of the unknown coefficients. It can be inverted in closed form, which is the basic idea behind the approach. The matrix $\mathbf{H}$ contains the unknown parameters in various combinations (p_{io}, $p_{io} + d_i$, $p_{io} - d_i$). The origin is temporarily moved to the present value of $\mathbf{p}$ by subtracting S_1. Following Lakanen (1976) then the inversion can be performed analytically:

$$S_{2i}' = S_{2i} - S_1$$
$$a_i = (S_{2i}' + S_{2i+1}')/2d_1^2.$$
$$a_{n+j} = (S_{2n+1+j}' - S_{21}' + S_{2k}')/d_1 d_k$$
$$a_{2n+i} = c_i = (S_{2i}' - S_{2i+1}')/2d_i$$

(4.33)

$$i = 1, 2, ..., n$$
$$j = (i^2 + i) (n^2 - n)/4 + k - i - 1$$
$$k = i+1, i+2, ..., n$$
$$l = 1, 2, ..., n-1 .$$

In order to find the extremum (minimum) of S the basic equation is derived with respect to each of the parameters. By rearranging, one gets an n x n matrix $\mathbf{B}$, the elements of which are:

$$b_{ii} = a_{2i}$$
$$b_{ii} = a_q \quad \text{when } i > j \qquad q = i \cdot n + j - (i^2 + i)/2 .$$
$$b_{ji} = b_{ij} , \quad i = j = 1, 2, ..., n$$

With $c_i = (S_{2i}' - S_{2i+1}')/2d_i$. [$i = 1, 2, ..., n$] the extremum of the hyperparabola is:

$$P_{extr} = B^{-1} (- C^T).$$

(4.34)

This completes one search round and the iteration is continued by:

$$P_{new} = P_{extr} + P_o.$$

(4.35)

If one wishes to be on the safe side, it still has to be checked that $S(P_{extr})$ truly is a minimum.

Lakanen (1976) has given the location of the minimum of the object function for the cases 1, 2 and 3 parameters in Table 4.2.

TABLE 4.2. Parameters and location of the extremum for hyperparabolic fit

#	N	P_{extr}
2	3	$p_1 = p_{1o} + -c_1/2a_1$
3	6	$p_1 = p_{1o} + (a_3 c_2 - 2a_2 c_1)/C$
		$p_2 = p_{2o} + (a_3 c_1 - 2a_1 c_2)/C$
		$C = 4a_1 a_2 - a_3^2$
4	10	$p_1 = p_{1o} + (2a_2a_6c_3 + 2a_3a_4c_2 - a_6^2c_1 - 4a_2a_3c_1 - a_4a_6c_3 - a_5a_6c_2)/D$
		$p_2 = p_{2o} + (2a_1a_6c_2 + 2a_3a_4c_1 + a_5^2c_2 - 4a_1a_3c_2 - a_4a_5c_3 - a_5a_6c_1)/D$
		$p_3 = p_{3o} + (2a_1a_6c_2 + 2a_2a_5c_1 + a_4^2c_3 - 4a_1a_2c_3 - a_4a_5c_2 - a_4a_6c_1)/D$
		$D = 8a_1a_2a_3 + 2a_4a_5a_6 - 2a_1a_6^2 - 2a_1a_5 - 2a_3a_4^2$

4.3.1.3 Pattern Search

The pattern search method of Hooke and Jeeves (1961) contains two basic search procedures, one along the original parameters axes (exploratory move) and a second one (pattern move) in the direction of greatest change (Fig. 4.17). These two procedures are applied alternately. As Fig. 4.18 shows, the pattern search technique adjusts itself well to follow strongly curved objective functions. In another classical method of non-linear optimization, the Rosenbrock method, the search is done in the direction of the original parameters only during the first complete iteration cycle. After that every new iteration cycle starts with the definition of a set of orthonormal, conjugate directions b_j, $j = 1$, 2,...M.

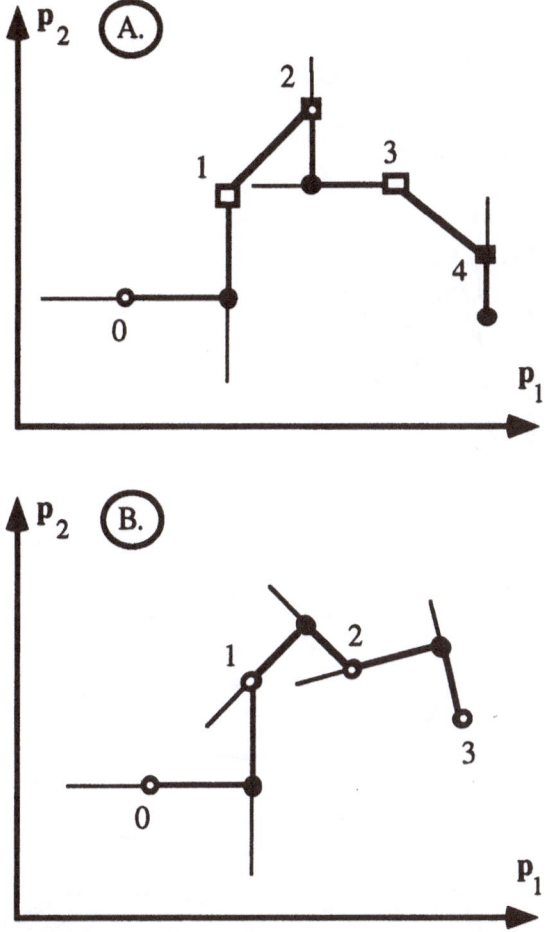

Fig. 4.17. Pattern search in the space of two parameters. **A.** The method of Hooke and Jeeves (1961). **Lines with circles:** exploratory move; **lines with squares:** pattern search move; **numbers:** iteration #; **thin line:** direction of unsuccessful search; **thick line:** direction of successful search. **B.** Conjugate direction method of Rosenbrock. **Circle 0:** start of search; **Circles 1, 2, 3, ...:** start of new iteration, determination of conjugate directions; **thin line:** direction of unsuccessful search; **thick line:** direction of successful search

The first component b_1 is always taken in the direction defined by the starting and ending points of an iteration cycle (cf. Fig. 4.17 B)

$$b_1^{(k+2)} = (p^{(k+1)} - p^{(k)})/\|p^{(k+1)} - p^{(k)}\|. \tag{4.36}$$

The other components of $\mathbf{b}$ are defined by $b_{j+1}^{(k+2)} \perp b_j^{(k+2)}$, $j = 2,3, ...M$, which leads to

$$b_j^{(k+2)} = b_j^{(k+2)} - \sum_{m}^{j-1} (a_m^{(k+1)T} b_m^{(k+1)}) b_m^{(k+1)} \tag{4.37}$$

with:

$$a_m^{(k+1)} = p^{(k+1)} - p^{(k)} - \sum_{m}^{J-1} (p_m^{(k+1)} - p_m^{(k)})$$

Finally, by normalization to unit length:

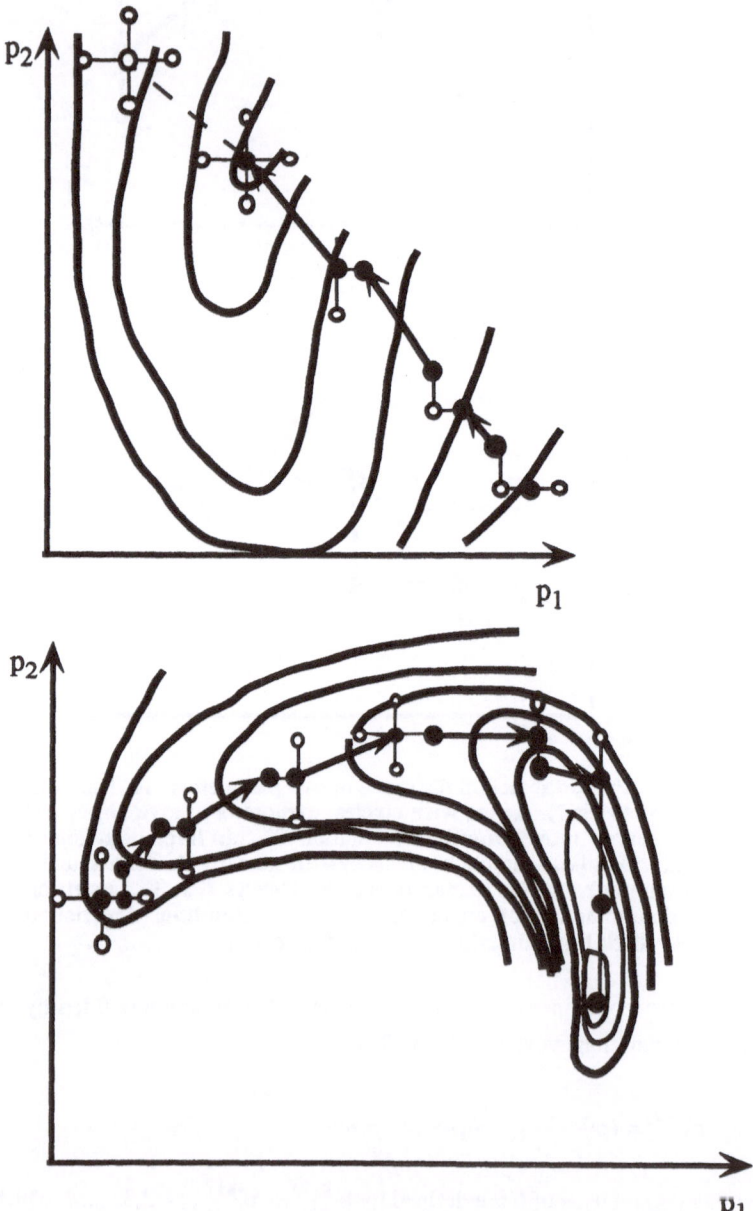

Fig. 4.18. Finding (**top**) and following (**bottom**) a ridge by pattern search. (Summarized from various books on optimization , e.g. Wilde 1964)

$$b_j^{(k+2)} = b_j^{(k+2)}/|b_j^{(k+2)}| \tag{4.38}$$

The Rosenbrock procedure is an efficient search method at the beginning stage of optimization, when the search is far from the optimum point of the objective function S.

4.3.2 Multivariate Search

Multivariate search usually refers to understood methods where the search for optimum is performed by exploring two or more parameters simultaneously. Using this definition in a strict sense the hyperparabolic fit above, the second procedure in the pattern search algorithm by Hooke and Jeeves and the Rosenbrock algorithm are multivariate in nature. In the following one multivariate method of the pattern search type, the simplex method, and some ideas behind the principle of steepest descent are presented. For further consultations. the books by Wilde (1964), Kowalik and Osborne (1968), Cooper and Steinberg (1970), Daniels (1978) and Powell (1981, 1982) are recommended.

4.3.2.1 The Simplex Method

In the simplex method a M-dimensional space is spanned by M+1 parameter vectors. The geometric figure formed by these points, the *simplex* has lent its name to the method. A two-dimensional simplex is thus a triangle (cf. Fig. 4.19), a three-dimensional simplex a tetrahedron and so forth. Let the initial value of the original parameter vector be p_o representing the point P_O in the simplex space. To start the simplex search, M additional points will be needed. These points can be selected arbitrarily, but they should cover the (M+1)-dimensional simplex space as smoothly as possible. Daniels (1978) suggests a selection, where each component of p_o is increased by 10% in turn. Thus:

$$P_i = (p_1, p_2, \dots 1.1 \ p_i, \dots, \ p_M) \qquad \text{for i = 1, 2, ..., M}.$$

The iterative part of the search consists of ordering the points according to the value of the objective function S. New simplex points are generated by three basic operations: reflections, contractions and expansions.

The first operation is to locate the *centroid* point (P_C) of the simplex, which is the average location of all points except the point with the highest S-value (assumed to be P_O in Fig. 4.19). A new simplex point is generated at R via *reflection* of point P_H with respect to the centroid using:

$$P_R = (1 + \alpha) \ P_C - \alpha \ P_H \qquad\qquad 1 \geq \alpha > 0. \tag{4.39}$$

A *value close to 1 is generally preferred*, but using exactly 1 is bound to create numerical difficulties. If $S(P_R) < S(P_H)$, then P_H is abandoned and P_R is incorporated as a new simplex point. A new reflection will be performed.

If the reflection were so successful that $S(P_R) < S(P_L)$, the lowest S-value among all the present simplex points, the algorithm assumes that further improvements can be expected in the same direction. An *expansion* is performed according to:

$$P_E = \beta\, P_R + (1 - \beta)\, P_C \qquad\qquad \text{with } \beta > 1. \qquad (4.40)$$

The point with smaller value of S [P_R or P_E] will replace P_H. A new centroid is determined and the search proceeds with another reflection.

If the first reflection was less successful and $S(P_R) > S(P_H)$, then the assumption is that a better (smaller) S-value can be found on the other side of the centroid. A *contraction* takes place:

$$P_X = (1 - \gamma)\, P_C + \gamma\, P_H \qquad\qquad \text{with } 0 < \gamma < 1 \qquad (4.41)$$

i.e. a search along the line P_H - P_C is performed. For a successful contraction P_X replaces P_H and a new centroid-reflection procedure takes place. If, on the contrary, $S(P_X) > S(P_H)$, the existing simplex has to be rescaled by:

$$P_i + k\,(P_L - P_i) \rightarrow P_i \qquad\qquad \text{for all } i = 0, 1, ..., M. \qquad (4.42)$$

If $0 < k < 1$, the simplex is shrinking around the point P_L. If $k < 0$, the simplex will increase in size. Daniels (1978) suggests numerical values 0.5 for reduction and -1 for increase.

This description of the simplex method followed the exposition of Daniels (1978), where descriptive numerical examples and a FORTRAN program for the method are also given. Another excellent description and geophysical example is presented by Whitehill (1973).

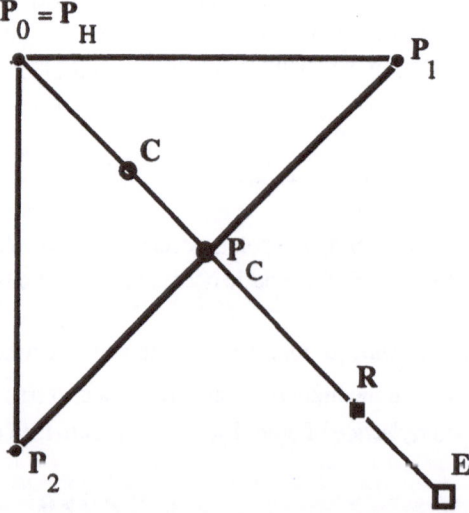

Fig. 4.19. The simplex method in search in the space of two parameters. The simplex consists of points **PO, P1** and **P2**. **PC**: the centroid point; **C**: point of contraction; **E**: point of expansion; **R**: point of reflection

4.3.2.2 Steepest Descent Methods: the Gradient Method

If the objective function is a reasonably smooth and well-behaved function, it seems obvious that a good strategy of optimization would be to always proceed in the direction of greatest change of S. The gradient vector at any point p_0, $g(p_0) = \nabla S(p_0)$ is perpendicular to the contour of S = constant passing through the point p_0. The negative gradient points in the direction of greatest rate of decrease of S. This is the backbone of the *method of steepest descent,* an old and popular technique dating back to the time of Cauchy (Cooper and Steinberg 1970).

For reasons described under the Newton 's method earlier, a steepest descent algorithm using only the gradient is sensitive to numerical errors. The literature on optimization methods is full of various techniques ranging from the use of higher order derivatives to correction matrices to update the direction of search and to stabilize the algorithm.

Each step of kth iteration of the basic steepest descent method consists of calculation of the gradient $g^{(k)}$ and of mapping in the direction of the negative gradient. Formally, the mapping can be written as a local minimization problem for the step λ. We have λ_k

$$\lambda_k = \text{Min } S(p^{(k)} - \lambda g^{(k)}) \tag{4.43}$$

and the starting point in the parameter space for the next iteration is

$$p^{(k+1)} = p^{(k)} - \lambda_k g^{(k)}. \tag{4.44}$$

The local minimization may be performed by any suitable of the univariate methods discussed earlier. Classification of the variants of the gradient methods (if necessary) can be based on whether the gradient is calculated analytically or numerically and based on the method used for the local λ-minimization. Instead of Eq. (4.43) the local linearization of S around $p^{(k)}$ can be used (Tarantola 1987).

The gradient method works well for smooth S-functions, when the gradient is easy to evaluate accurately and when the changes in the various parameters are of equal size. For a quadratic S with circular contours of S = constant, the method actually finds the minimum in a single iteration (Fig. 4.20 A). For elliptic contours, the gradient method quickly slows down when the minimum is approached (Fig. 4.20 B). The numerical instability of the method, described earlier in the attractor framework, can be visualized in the optimization framework as shown in Fig. 4.20 C. If for reasons of numerical accuracy, the gradient is calculated for a point with $p + \Delta p$, instead of for the original point (with p), the direction of the gradient may become erroneous in a disastrous way. This sensitivity of the gradient method to numerical errors was described in Section 4.2.5 in connection with the one-dimensional Newton method.

4.3.2.3 Steepest Descent Methods: the Method of Conjugate Directions.

Let us consider a modification of the local univariate search for the gradient method:

$$\lambda_k = \text{Min } S\left(p^{(k)} + \lambda\, v^{(k)}\right). \tag{4.45}$$

The directions v are called conjugate directions, if:

$$v_j^T \underline{A}\, v_i = 0. \tag{4.46}$$

They form new directions of search for the optimum of S, directions which are linearly independent. If S is quadratic with respect to the positive definite matrix $\underline{A}$, then the minimum of S is found by searching in each of the directions v_i only once. The steepest descent method works, in contrast to the gradient technique, best close to the optimum value (Fig. 4.21). Various constructions of the vector set v will give different optimization algorithms. Let us briefly mention a few of them.

The class of CG (conjugate gradient) methods consists of those for which iterates of the parameter vector are generated according to the rules:

$$p_i = p_{i-1} + \alpha_i\, v_i$$
$$\beta = \frac{\frac{1}{2} g_i^T H y_i}{v_i^T y_i} \tag{4.47}$$
$$v_{i+1} = -H\, g_i + \beta_i v_i$$

Here, p_0 is given and:

$$g(p) = \nabla S(p)$$
$$g_i = g(p_i)$$
$$v_1 = -H g_o \tag{4.48}$$
$$y_i = g_i - g_{i-1}$$

Also, α_i is a suitably chosen scalar and H a known positive definite matrix. When $H = I$, the algorithm becomes the original conjugate gradient algorithm of Fletcher and Reeves (1964).

For quadratic S, β_i may alternatively be given by:

$$\beta = \frac{g_i^T H\, g_i}{g_{i-1}^T H\, g_{i-1}}\,,$$

or:

$$\beta = \frac{g_i^T H\, y_i}{g_{i-1}^T H\, g_{i-1}}\,, \tag{4.49}$$

but the form given first is most suitable for many purposes (Powell 1982).

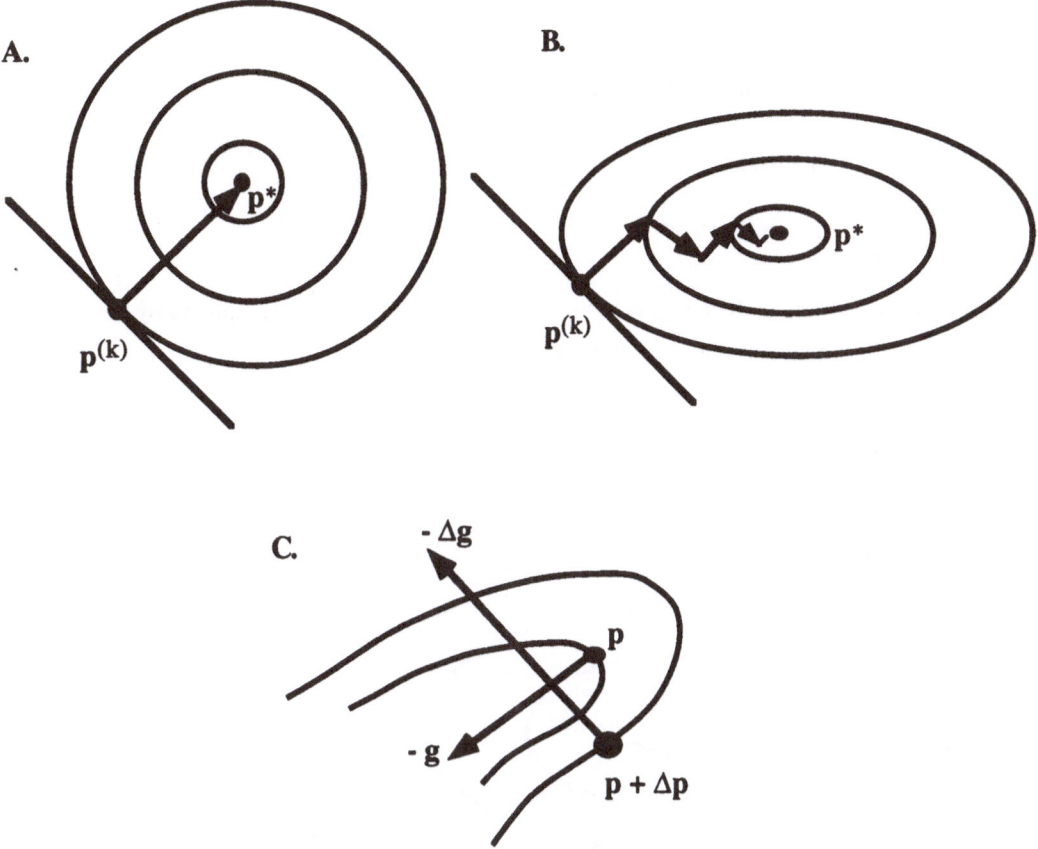

Fig. 4.20. Behaviour of the gradient (steepest descent) method. **A.** The optimum is found in one step, if S = const are circular; **B.** the method slows down, if S = const are elliptical; **C.** the method is sensitive to numerical errors. (summarized from various books on optimization, e.g. Wilde 1964)

Another class of descent methods takes into account higher order derivatives of the original objective function S $(\mathbf{p}^{(k)} + \lambda \, \mathbf{v}^{(k)})$. Taking the terms up to the second degree:

$$\frac{\partial S}{\partial \lambda} = 0 = \mathbf{v}^T \cdot \nabla S \, (\mathbf{p} + \lambda \, \mathbf{v}) \approx \mathbf{v}^T \cdot [\nabla S \, (\mathbf{p}) + \lambda \, \mathbf{J} \, \mathbf{v} + ...) \, , \tag{4.50}$$

where $\mathbf{J} = \left\{\dfrac{\partial^2 S}{\partial p_i \partial p_j}\right\}$ is the Jacobian of the objective function. For many geophysical models $\mathbf{J}$ can be suitably determined analytically. Otherwise again various approximations to $\mathbf{J}$ result in different variants of optimization algorithms.

One very widely used optimization method of this category was developed by Powell in 1971 (Powell 1981, 1982). Its popularity stems from two features at least. It is a very stable (albeit sometimes slow) method and no derivatives of the objective function are needed. The Powell algorithm can be described as follows:

1. $\qquad \lambda_0 = \text{Min} [S(\mathbf{p}^{(1)} = \mathbf{p}^{(0)} + \lambda \, \mathbf{v}_M)]$

2. $\qquad \lambda_k = \text{Min} [S(\mathbf{p}^{(k+1)} = \mathbf{p}^{(k)} + \lambda \, \mathbf{v}_k)] \qquad\qquad k = 1, 2, ..., M$

3. $\qquad \mathbf{v}_i = \mathbf{v}_{i+1} \qquad\qquad\qquad\qquad\qquad\qquad\qquad i = 1, 2, ..., M\text{-}1$

4. $\qquad \mathbf{v}_M = \mathbf{p}^{(M+1)} - \mathbf{p}^{(M)} \, .$

If the stopping criteria of the iteration have not been fulfilled then $\mathbf{p}^{(o)}$ is replaced by the last value of $\mathbf{p}^{(M+1)}$ and the next iteration starts again from 1.

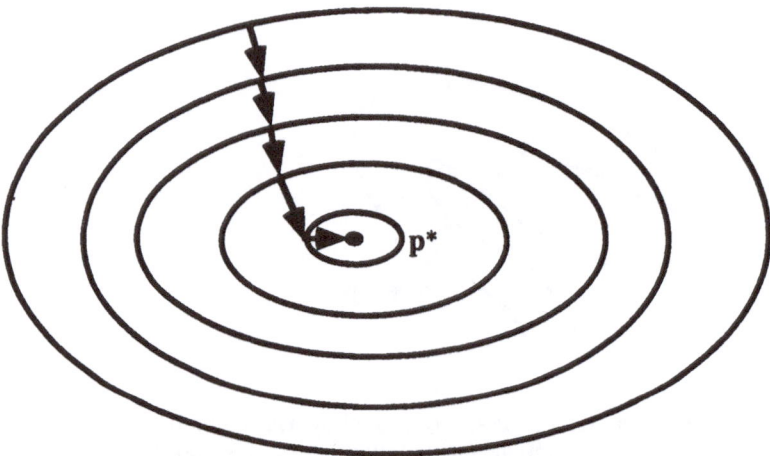

Fig. 4.21. Behaviour of the steepest descent method for elliptical contours of S. (After Daniels 1978)

4.3.2.4   The Levenberg-Morrison-Marquardt Algorithm

One optimization method which has won a great popularity deserves especially to be mentioned, the more so, since it has intimate connections both with the gradient method and with the generalized linear inversion discussed earlier. The method was for the first presented by Levenberg (1944) but was soon forgotten probably because of the war. Later, Morrison reinvented the algorithm and Marquardt (1963) finalized it into its present form.

If we take the first order term the change of the model parameter vector, $\Delta\mathbf{p}$, one iteration step of the linearized inversion consists of the solution of the linear system of equations:

$$\underline{\mathbf{M}}^{(k)} \, \Delta\mathbf{p}^{(k)} = \underline{\mathbf{G}}^{(k)^T} \mathbf{F}_h^{(k)} , \qquad\qquad\qquad\qquad (4.51)$$

with $\qquad \underline{M}^{(k)} = \underline{G}^{(k)^T} \underline{G}^{(k)}$,

and $\qquad \underline{G}^{(k)} = \left\{ \dfrac{\partial F_{h\,i}}{\partial p_j} \right\}$.

The solution of the system of equation can be problematic in case the matrix $\underline{M}$ is ill-conditioned (as discussed in Chap. 3). In order to enforce the diagonality of the matrix, the (Levenberg-Morrison-) Marquardt algorithm uses an additional parameter α:

$$\underline{M}^{(k)} : \underline{M} + \alpha^{(k)}\,\underline{I} \ . \qquad\qquad (4.52)$$

Marquardt introduced normalization in the form:

$$\underline{M}_{norm} = \left\{ M_{p,q}/\sqrt{M_{pp}M_{qq}} \right\}, \qquad\qquad (4.53)$$

$$C_{norm,\ p} = C_p/\sqrt{M_{pp}} \qquad\qquad \text{with } C = \ \underline{G}^T F_h$$

and $\Delta p_{norm,\ q} = \Delta p_q \cdot \sqrt{M_{qq}}$

The additional parameter α can be used as an efficient alternative to dropping of the zero or small eigenvalues in the SVD linearized approach of Section 3.3. The proper selecting of α from one iteration to another depends on the model used, on the properties of the objective function with respect to the parameters and requires often quite a lot of experimenting. (In fact, Levenberg used in his proposition of the method different parameters for each of the parameters, using the diagonal matrix $\underline{A}$ instead of $\alpha\underline{I}$.) One can, however take advantage of the fact that when $\alpha \to 0$ the algorithm actually makes an iteration in the direction of the gradient of S (= Newton direction). When $\alpha \to \infty$, the direction of steepest descent is approached. Thus the magnitude of α should change so that the gradient direction should dominate when being far from the optimum and the steepest descent direction close to the optimum.

4.3.3   Random Search Methods

Recently several statistical interpretation procedures have been described and gained rather wide popularity. In these methods the search for optimal parameter values is completely random, the classical ones were called Monte-Carlo methods. The parameters are given values completely stochastically (albeit within given limits and often also with an assumed statistical distribution). The model response and the objective function is calculated for each of the parameter combinations and all those values are retained which give a response fitting the data with a prescribed accuracy. This approach normally requires a huge number of response calculations in order to give some reasonable idea about the average (or median) value of the parameters. This allows for some statistics to give an idea about the resolution of the parameters. However, the method is not in itself a cure for the unambiguity problems. This is well demonstrated by one example taken from a paper of Jones and Hutton (1979), where several single-station magnetotelluric soundings were inverted by the Monte-Carlo technique. The 1-D three-layer model did allow a great number of response function calculations. All models with response curves fitting the data and their error bars within the acceptance level were added to the results of the model data base.

The envelope of all acceptable models for one of the stations is shown in Fig. 4.22. Values of some selected parameters are given in the associated table for two acceptance levels. No greater differences are noted. The table shows that not even the traditionally well-solved parameters of magnetotellurics are very well determined in this case. The thickness of the topmost layer = depth to the surface of the conducting layer and the conductances of the two first layers have very great ranges of variation. The lines with arrows indicate the statistically possible locations of the transition between the layers. In fact, the overlapping of the lower value of the layer 1-2 boundary and the upper value of the layer 2-3 transition puts some doubt on how reliably the conducting layer is resolved in this inversion.

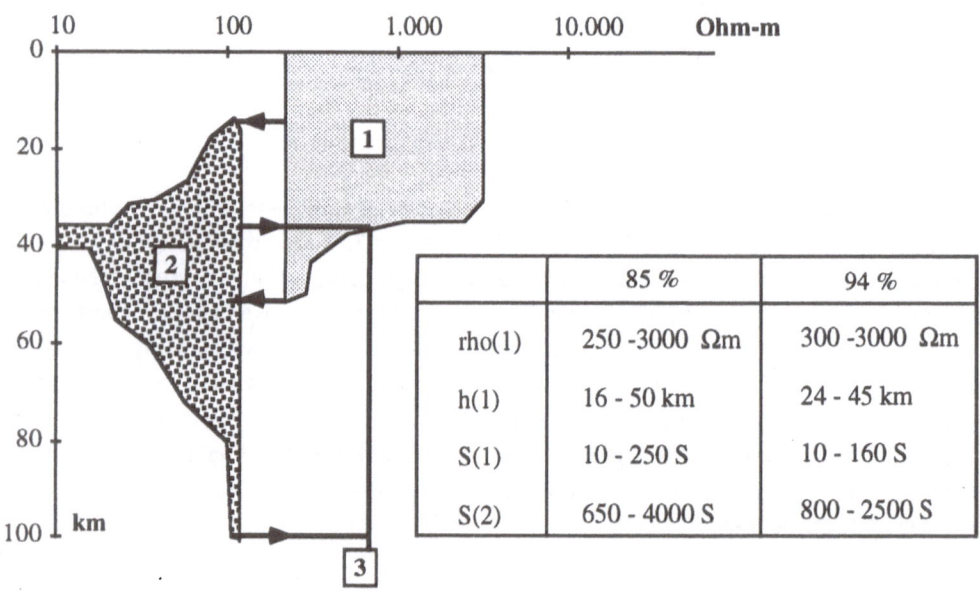

	85 %	94 %
rho(1)	250 -3000 Ωm	300 -3000 Ωm
h(1)	16 - 50 km	24 - 45 km
S(1)	10 - 250 S	10 - 160 S
S(2)	650 - 4000 S	800 - 2500 S

Fig. 4.22. Resistivity-depth profiles of acceptable models from Monte-Carlo inversion of single station magnetotelluric sounding. Only the smoothed outer boundary of the three-layer model is shown. The range of the selected model parameter is shown for 85 and 94 % acceptance levels. (Schematically adapted from Jones and Hutton 1979)

Most reports on the use of random methods indicate problems with computing time even for a modest number of parameters (see Mora 1988) and those approaches are recommended only in cases when the calculation of the forward problem is inexpensive (Tarantola 1987). Despite this some early successful geophysical applications of Monte-Carlo inversion can be found in the literature (Press 1968; Wiggins 1969; Press 1971). It is, however, possible that massively parallel computers of the future may make the situation more favourable for random search methods.

Many of the statistical algorithms make use of an analogy with a physical, chemical or biological process which appears in nature. The algorithms have often strongly heuristic features. One of these random search methods, called *simulated annealing* (Kirkpatrick et al. 1983; Aarts and Korst 1989), has been applied to geophysical inversion problems (Keilis-Borok and Yanovskaya 1967; Anderssen and Seneta 1971, 1972; Rothman 1985, 1986; Tarantola 1987; Jakobsen et al. 1988). In this approach the stochastic search is controlled by a method which is analogous to the physics that occurs when crystals are ordered in a metal when it is cooled during an annealing process. The objective function and total energy of a system are equivalents, whereas the different minima correspond to various energy states of the system. It seems that too few comparisons with other methods of optimization exist in order to decide, whether this approach could overcome the computer cost limitations of random search methods.

Genetic algorithms (Goldberg 1989; Davidor, 1990) are based on simplified models of population genetics and *tunneling algorithms*(Levy and Montalvo 1985) on the physical phenomenon of electron tunneling to mention but a few. These ideas remind us very much of the analogue modelling used in times before the computers, when physical fields based on the same mathematical equations were used to model the fields. So the calculation of electron trajectories in vacuum tubes was simulated by rolling metal spheres on a tightened rubber mat and by photographing the paths. Resistor networks simulated potential fields conveniently. Now the analogy of both phenomena to be simulated are performed numerically and in the computer. Both genetic and tunneling algorithms are still very much in a developing stage and they seem to suffer from the main disadvantage of stochastic methods, i.e. long computing time. Both algorithms are, however, intrinsically parallel in nature, so interesting developments are to be expected along with the improvements in parallel computers.

One of the newer stochastic algorithms has been coined the name *tabu search*. In fact, it is a higher level method or metastrategy of optimization, in which traditional optimization algorithms are embedded (see Glover 1989, 1990). The approach is strongly heuristic, and is designed so as to prevent the traditional search procedure from becoming trapped in a local minimum. Tabu search implements moves to transform one local optimal solution to another and memorization of the paths of the basic search procedure. Intensifying (= reinforcement of earlier discovered "good" regions) and diversification (= move into new areas of search) are controlled heuristically in the tabu search algorithm. Besides massively parallel computer systems, the use of neural networks (Hopfield 1985) seems to be in the forefront of research for certain, albeit not yet geophysical, optimization problems.

4.4 EXAMPLES

There is no general rule for choosing a method to be applied in non-linear optimization of geophysical inversion problems. Not too many comparative studies exist in the geophysical literature on the use of different optimization techniques for inverting data for a single geophysical method. Most studies (Al-Chalabi 1970, 1971, 1972; Bott 1973; Hjelt 1973, 1975) reach the conclusion that a *variable strategy* of optimization is to be preferred, the use of interactive computer terminals being of great help. During initial stages of the optimization, when the search is far away from the optimum, course methods like stochastic (Monte-Carlo) search should be chosen (Bott 1973). Close to the optimum, gradient methods are preferable (Bott 1973). According to Al-Chalabi, Simplex-type search is suitable for multimodel functions and Marquardt-type search for trivial functions. The number of parameters can be kept small by the technique of partial anomalies (Hjelt 1973, 1975), a technique especially prone to parallel computer implementations.

In the following a collection of examples from the geophysical literature are surveyed, a more thorough collection being taken from 2D magnetometry inversion, because of the familiarity of the author with this problem.

4.4.1   Magnetic 2D Profiling

The model of 2D magnetic profile interpretation example is a thick plate, with finite depth extent and a regular (parallelogram) cross-section. The calculation of the theoretical field of such a model has been discussed already in the Introduction and the properties of the objective function in LSQ fit in Section 4.1. Although a great number of optimization methods were tested, the simple direct search with parabolic speedup was chosen for the final computer program package developed (Hjelt et al. 1977). The main reasons were the requirements of interactivity and the efficiency of the partial anomaly concept, which kept the number of parameters reasonably low. Later, the multidimensional hyperparabolic search, 3D magnetic and both 2D and 3D gravity modelling facilities were added to the program package. Three- dimensional modelling was feasible only in single exceptional cases because of great computer time demands.

The background for the magnetic profile package was a thorough experimental study of the four major items of computerized geophysical inversion (Hjelt 1973):

1. Choice of interpretation method;
2. Choice of basic interpretation model;
3. Choice of starting values for the model parameters;
4. Iterative improvement of the parameter values.

The 2D magnetic thick plate model became popular through one of the first papers (Gay 1963) emphasizing the importance of using the whole geophysical profile for interpretation instead of single special points of an anomaly only. The model also seemed well suited to describe the geological structures of Finnish Precambrian bedrock. Six parameters were used to describe the model, since strike direction of the plate and remanent magnetization were considered known. In some experiments a first order polynomial was added to describe the regional gradient field along the measured profile. The gradient term in particular seemed to introduce problems of convergence, because of its correlation with the effects of the plate dip. Experiments to improve convergence by linearization of the plate dip parameter failed in most cases.

The sum of squares of the deviation between model field and measured data was the obvious choice for objective function, since tests with L_1 norm (maximum deviation) or weighted (using anomaly amplitude) sum of squares failed to give reliable performance or added non-desired features to the objective function contours (see Fig. 4.4). The starting values of the model parameters could be derived by the program automatically, but when the use of the program package became more interactive, the user-provided starting values more often than not turned out to result in quicker fit.

The computer code for five optimization methods were available for the rather extensive tests performed: in addition to the direct search with parabolic speedup (called sequential parabolic search by Hjelt 1973), the linearized method of Levenberg-Morrison-Marquardt, the conjugate direction methods of Powell (only model field values needed) and Davidon (both field and derivatives needed) and the (at that time) recently developed method called SPIRAL (Jones 1970). The last method was very promising in its approach, in which the parameter search was conducted along a spiral in the plane of the Newton and steepest descent directions (Fig. 4.23), turned out to be a disappointment (Hjelt 1975). The Levenberg-Morrison-Marquardt algorithm will be called, for short, the Marquardt algorithm as has become common practice in the recent geophysical literature.

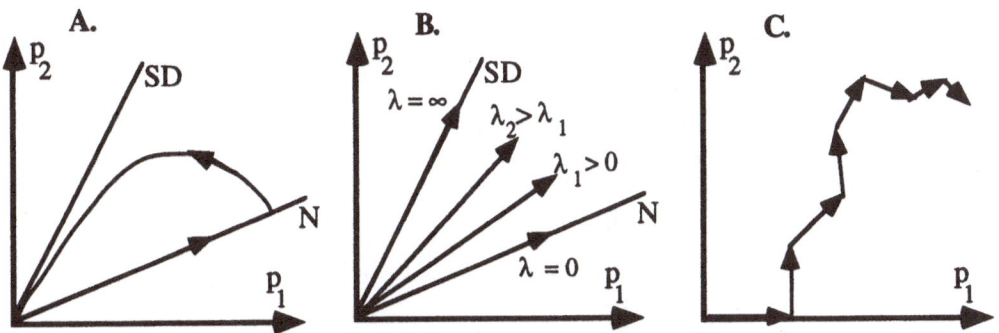

Fig. 4.23. Sketches of paths of parameter search in some non-linear optimization methods used in studies of 2-D magnetic profile interpretation. **A.** The SPIRAL method; **B.** the Levenberg-Morrison-Marquardt algorithm; **C.** the method of conjugate directions. **N** The Newton direction, **SD** direction of steepest descent [After from Hjelt 1973].

Since the aim of the program design was to handle long magnetic anomaly profiles, it was reasonable to introduce the concept of partial anomalies. Trying to fit the anomaly of a single structure described by a plate to the whole long profile produced very slow convergence rates for the iteration, almost irrespective of the method used. The concept should be of value, if implementing inversion programs for parallel computers. This idea would be less well suited, if the anomaly of the basic model is of maximum-minimum type. This would be the case for horizontally lying magnetized plates in high magnetic latitudes and vertical magnetized structures in equatorial latitudes.

In all the described examples the performance of the optimization method is described by plotting the objective function S (the sum of squares of errors of fit) as a function of number of function evaluations required. The amount of calculation of derivatives, which involved different functions than the anomaly field itself, was taken into account. The components of the parameter vector are ordered similarly in all cases, the plate parameters as $p_{plate} = (d = \text{thickness}, x_0 = \text{horizontal location of the upper surface}, z_0 \text{ depth to the upper surface}, h = \text{depth extent of the plate}, \Phi = \text{dip}, k = \text{susceptibility}; \text{see Fig. 4.24})$ and the additional parameters (if used) as $p_{add} = (\alpha = \text{the strike direction}, ZH = \text{regional background}, GR = ZH + b \ x = \text{regional gradient})$.

In the first example (Fig. 4.25) the "data" consist of 11 theoretical vertical component magnetic field values calculated for a plate model: $p_{plate} = (d = 30 \text{ m}, x_0 = 50 \text{ m}, z_0 = 10 \text{ m}, h = \text{fixed} =$

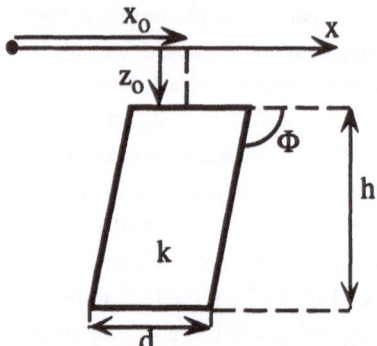

Fig. 4.24. The magnetized 2D plate. Parameters: d = thickness; x_0 = horizontal location of the upper surface; z_0 = depth to the upper surface; h = depth extent of the plate; Φ = dip; k = susceptibility

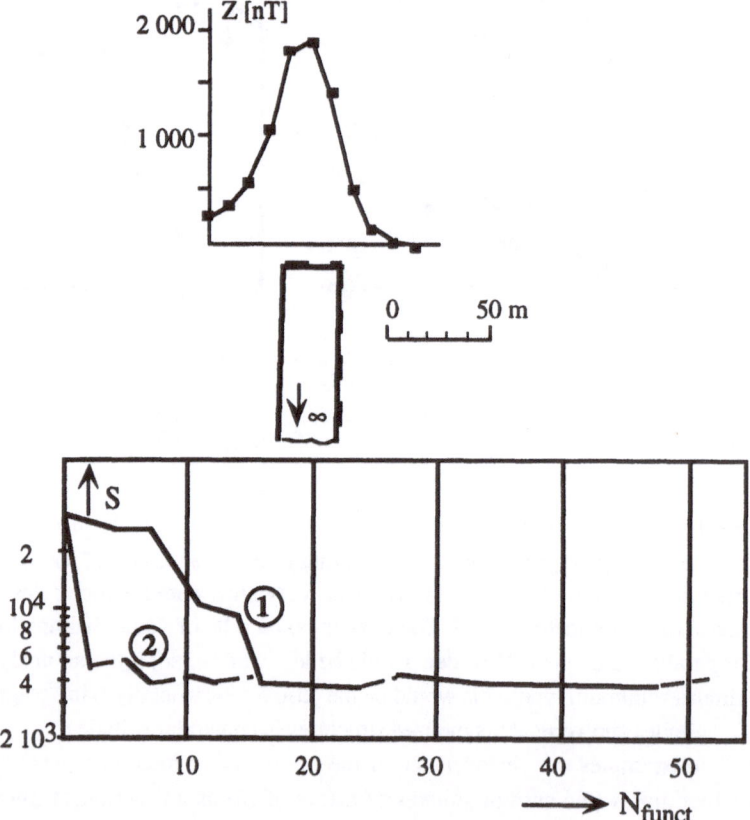

Fig. 4.25. Comparison of convergence of minimization between Marquardt and Powell algorithms for a theoretical single plate magnetic anomaly. **Upper part:** anomaly profiles. **Dots:** original model; **Line:** inverted model. **Middle part:** models. Dashed lines original; continuous lines inverted, p_{inv} = (d = 27.4 m, x_0 = 49.6 m, z_0 = 11.1 m, h = fixed = ∞, Φ = 90.6 °, k = 0.0114 cgs units). **Lower part:** convergence of inversion. **1.** Powell; **2.** Marquardt. **Dotted lines:** extra dip linearization. (After Hjelt 1973)

∞, $\Phi = 90^0$, k = 0.01 cgs units) and p_{add} = (α = fixed, ZH = fixed = 0, GR = 0). The parameters are recovered by optimization to an accuracy better than 10% on average. The obtained level of fit has a variance slightly above 1 % of the maximum amplitude of the anomaly. The linearized Marquardt algorithm proceeds faster to the final level of fit than the Powell algorithm (Fig. 4.25) The dotted parts of the convergence curves relate to the attempts to use an additional linearization algorithm for the dip and susceptibility determination. It is evident that it disturbs the systematics of the optimization algorithms and therefore does not improve the inversion. The disturbance is larger in the Marquardt algorithm, which is understandable, since the algorithm itself is based on linearization.

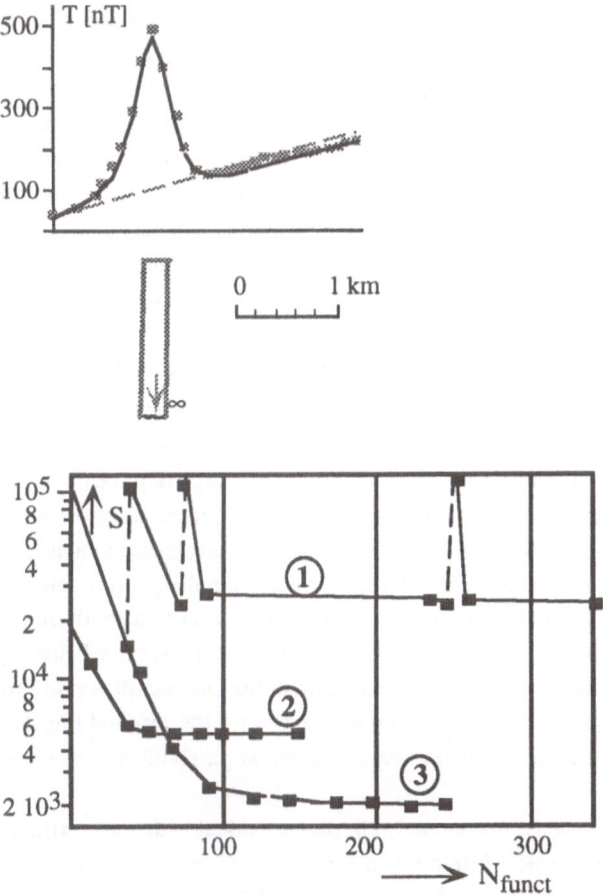

Fig. 4.26. Comparison of convergence of the Powell algorithm when using various parameter combinations in inverting a theoretical airborne single plate, total field anomaly. **Upper part:** anomaly profiles. **Dots** original model; **Line** inverted model. **Middle part:** inverted model, p_{inv} = (d = 261 m, z_0 = 150 m = fixed, h = fixed = ∞, Φ = 93 ^{0}, k = 0.0029 cgs units). **Lower part:** convergence of inversion. **1.** (d, x, h, GR) with Powell; Φ and k with dip linearization; **2.** (d, x, h, Φ and k) with Powell, GR = const.; **3.** (d, x, h, Φ, k and GR) with Powell. **Squares:** end of each iteration cycle; **dotted lines:** dip linearization. (After Hjelt 1973)

When a linear gradient was added to the plate anomaly simulating the regional gradient of a airborne magnetic measurement, the dip-linearization procedure was especially devastating in that it systematically counteracted the optimizing effect of each Powell iteration (Fig. 4.26). If the gradient was estimated before inversion and kept constant the optimization proceeded quickly to a minimum, which was of local nature. This is obvious, since adding GR to the parameters optimized by the algorithm produced a value about half a decade better for the objective function. The number of function evaluations was considerably large for this example with $p_{plate} = (d, x_o, z_o = fixed, h, \phi_i, k_i)$ and $p_{add} = (\alpha = fixed, ZH, GR)$.

In the next example real field data have been used (Fig. 4.27). The effect of two distinct magnetized bodies is evident on the profile. We have $p_{plate} = (d_i, x_{oi}, z_{oi}, h_i = fixed = \infty, \phi_i, k_i); i = 1, 2$ and $p_{add} = (\alpha, ZH, GR) = fixed$. The performance of the various algorithms almost follows theoretical expectations: both conjugate gradient techniques behave similarly, SPIRAL is slightly more effective and Marquardt the most efficient. The objective function is obviously close to unimodal and fully convex. Direct search with parabolic fit compares favourably with the fastest method. This is not due to differences in starting values of the parameters, nor to inclusion of the dip among the optimized parameters, but to the use of the partial anomaly concept. Since the anomalies of both plates are distinct, fitting of the parameter of either of the plates can be made to the essential part of the anomaly profile, thus speeding up the convergence. The variance of the model fit is about 2% of the maximum amplitude along the profile.

The fourth example is a measured multiplate aeromagnetic total field anomaly (Fig. 4.28). Four distinct anomaly peaks can be identified. Here $p_{plate} = (d_i, x_{oi}, z_{oi} = fixed, h_i = fixed = \infty, \phi_i, k_i); i = 1, ..., 4$ and $p_{add} = (\alpha, ZH, GR) = fixed$. The fit shown (variance > 5% of the maximum amplitude) is for the version, where all plates were supposed to have a large depth extent. This clearly is not true for the fourth plate (on the right), and has resulted in a poor determination of its dip (it is clearly the wrong direction). It seems probable that measuring errors have caused the inversion program (which was allowed to run automatically without intervention from the geophysicist) to get stuck at a local minimum. The depth to the upper surface has been kept constant (= flight height 150 m) for the convergence curves. Similar experiments with variable z_0 give the same behaviour and the same final level of the objective function for the Powell method; the Marquardt method for this case performs comparably to the sequential parabolic search for fixed z_0 (cf. original Fig. 12 in Hjelt 1973). It is evident again that the Marquardt method and sequential parabolic search outpowers the conjugate gradient methods.

The advantage of the partial anomaly concept is evident in the last, extraordinary example (Fig. 4.29). A 1.5-km-long profile of data taken by a vertical component ground magnetometer was inverted automatically, starting with the identification of the anomaly maxima and initial values of the parameters. Although there are considerable discrepancies in the fit, the general result is remarkable, considering the complexity of the model. The horizontal locations of the plates were not constrained which resulted in some plates partly overlap (e.g. plates 13 and 14, indicated by dashed lines of the model in the Fig. 4.29). For the overlapping parts, the susceptibilities of the plates can simply be added, since the k-values are small and no demagnetization or other non-linear effects need to be taken into account.

For this example $p_{plate} = (d_i, x_{oi}, z_{oi}, h_i, \phi_i = fixed, k_i); i = 1, 2,...., 18$ and $p_{add} = (\alpha, ZH, GR) = fixed$. The dip of all plates was assumed constant along the whole profile. The automatic

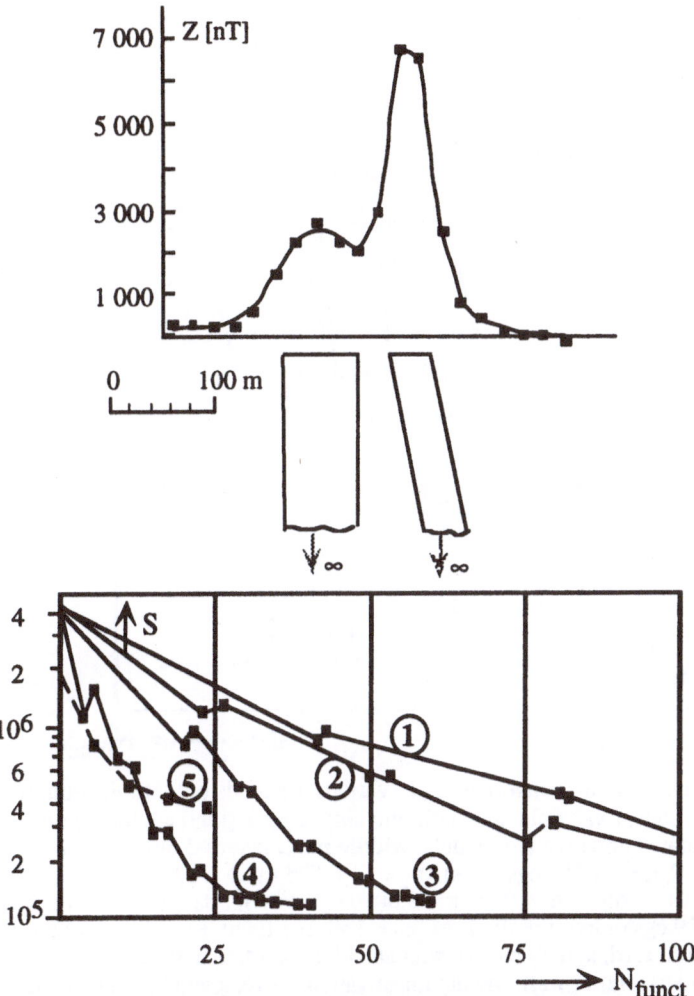

Fig. 4.27. The convergence of various optimization methods applied to inverting measured, two-plate vertical field magnetic anomaly. **Upper part:** anomaly profiles. **Dots** data; **Line** inverted model. **Middle part:** inverted model, $p_1 = (d = 72.8$ m, $z_0 = 18.0$ m, h = fixed = ∞, $\Phi = 88.5°$, k = 0.0105 cgs units), $p_2 = (d = 41.2$ m, $z_0 = 14.6$ m, h = fixed = ∞, $\Phi = 78.2°$, k = 0.0379 cgs units). **Lower part:** convergence of inversion, all with dip linearization except the direct parabolic search, for the algorithms: **1.** Powell; **2.** Davidon; **3.** SPIRAL; **4.** Marquardt **5.** direct parabolic search. **Squares:** end of each iteration cycle; **dotted lines:** dip linearization. (After Hjelt 1973)

inversion was repeated for four or five values of dip, the $45°$ shown giving by far the smallest objective function and best fit. It also happened, that this value corresponded to the average geological dip of the region of measurement. This successful example would be quite an achievement for any optimization technique: 18 plates with 5 parameters each means that the total number of model parameters amounted to 90! Five iterations, requiring 150 function evaluations (= 1 complete profile) for each dip value, required only a reasonable computer CPU time [a few minutes on the IBM 360/40 computer of Outokumpu Co (this was during the first part of the 1970s!)].

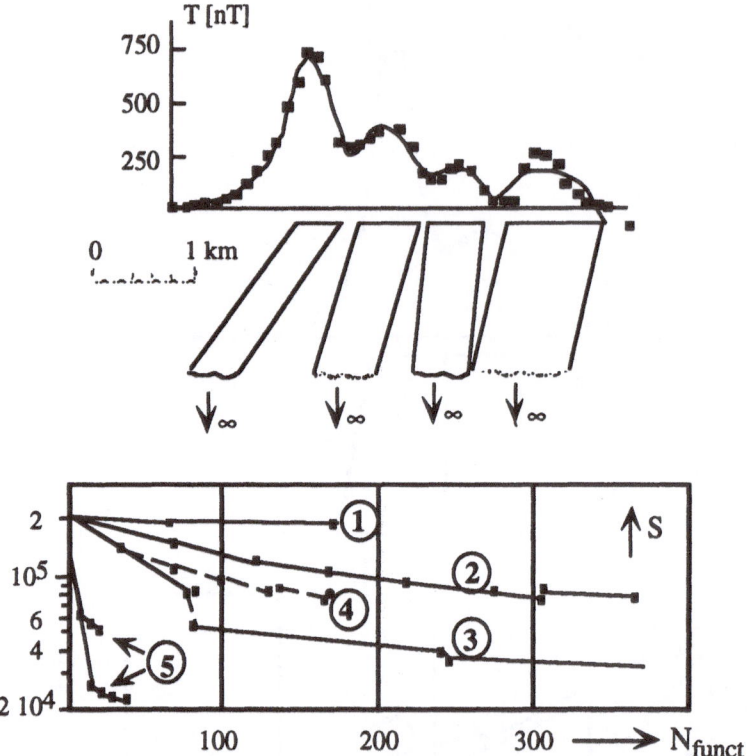

Fig. 4.28. Comparison of convergence of various methods of optimization for measured, multi-plate, total field magnetic anomaly. **Upper part:** anomaly profiles. **Dots** original data; **Line** inverted model. **Middle part:** inverted model, p_1 = (d = 480 m, Φ = 126 ^{O}, k = 0.0052 cgs units), p_2 = (d = 605 m, Φ = 107.5 ^{O}, k = 0.0031 cgs units). p_3 = (d = 565 m, Φ = 97 ^{O}, k = 0.0010 cgs units), , p_4 = (d = 917 m, Φ = 103 ^{O}, k = 0.0019 cgs units), for all plates: z_0 = 150 m = fixed. **Lower part:** convergence of inversion. **1.** (d, x, h, Φ and k) with Davidon; **2.** (d, x, h, Φ and k) with Powell; **3.** (d, x and h) with Powell and dip linearization; **4.** (d, x and h) with Davidon and dip linearization; **5.** (d, x, h, Φ and k) with direct parabolic search. **Squares:** end of each iteration cycle; **dotted lines:** dip linearization. (After Hjelt 1973)

Some additional details on the sequential parabolic search algorithm and relative computer times needed for one minimum iteration of various optimization used in the examples are given in Tables 4.3 - 4.5. The general conclusions of these and a great number of other experiments are as follows (Hjelt 1973, 1975):

1. The computer code used for the Powell algorithm was reliable for the greatest number of different parameter combinations, but slow when the number of plates (parameters) increased.
2. The linearized Marquardt algorithm was fast and reliable for profiles with small number of plate models (N_{plate} = 1...3).
3. Sequential parabolic search combined with partial anomalies is most suitable for interactive inversion and when N_{plate} > 2 (and probably also for parallel computer implementations)
4. For long profiles and complicated anomalies, dip and depth extent should be constant and optimization performed for a selected number of values for these parameters.

These findings do not differ essentially from results reported by other authors (see Al-Chalabi 1970; McGrath and Hood 1970, 1973; Bott 1973). An example of the influence of the Marquardt optimization parameter lambda on convergence in 2D multiplate magnetic inversion is given in Fig. 4.30

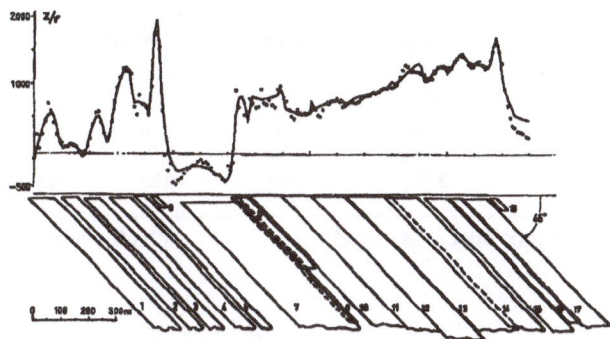

Fig. 4.29. Inverting a multiplate, vertical field magnetic anomaly using sequential parabolic search of partial anomalies. Constant dip, strike direction and regional field were assumed. The result shown is for best fitting dip (after Hjelt 1973). **Upper part:** anomaly profiles. **Dots** original data; **Lines** anomaly of inverted model. **Lower part:** inverted model

TABLE 4.3. List of plate parameters for the magnetic inversion example of Fig. 4.29 (Hjelt 1973)

Plate	1	2	3	4	5	6	7	8	9
d/m	95	59	63	76	45	25	227	33.5	57
z_0/m	18	14.6	20.6	14.4	7.6	10.7	32.5	15	10.9
k/cgs	0.00727	0.00315	0.00744	0.00922	0.00378	0.01837	0.00142	0.00463	0.00585

Plate	10	11	12	13	14	15	16	17	18
d/m	122	112	123	168	114	81	49	94	26
z_0/m	1.5	3.6	7	7.6	11.4	12.6	12.5	15.5	13.6
k/cgs	0.00511	0.00381	0.00481	0.00500	0.00609	0.00543	0.00570	0.00417	0.00950

TABLE 4.4. Features of sequential parabolic search (= direct search with parabolic speedup) for 2D magnetic profile inversion using plate models (Hjelt 1975)

Para-meter	Typical range	Limits of parameter Lower	Upper	Starting value	Step of search during 1st iteration
d	10 ... 400 m	5 m	1 000 m	From anomlay half-width	0.1 d_1
x_0	x_1 .... x_N	x_1 - 200 m	x_N + 200 m	From anomaly maximum	dx
z_0	> 0.2 m^a	0.2 m^a	100 m^a	Given or 10 m^a	5 m
h	100 ... 2 500 m	10 m	10 000 m	Given or 5 000 m	- 0.4 d
ϕ	45° .... 135°	0.1 rad (~ 6°)	3 rad (~172°)	Given	10°
k	0.005 .... 0.5 SI	0.001 SI	1.2 SI	From anomaly max amplitude	0.25 k

^{a} Below the Earth's surface.

TABLE 4.5.   Selected normalized minimum computing times for an iteration cycle using different optimization algorithms (After Hjelt (1975)).

# Plates	# Data [a]	F_{theor}	Marquardt	Dir.search	Powell
1	10	1 [b]	2	24	12
2	20	3,5	8	50	89
5	50	22	48	140	135
20	200	355	760	820	85 000

[a] # of data = 10 times # of plates (assumed).   [b] Reference (normalized to 1)

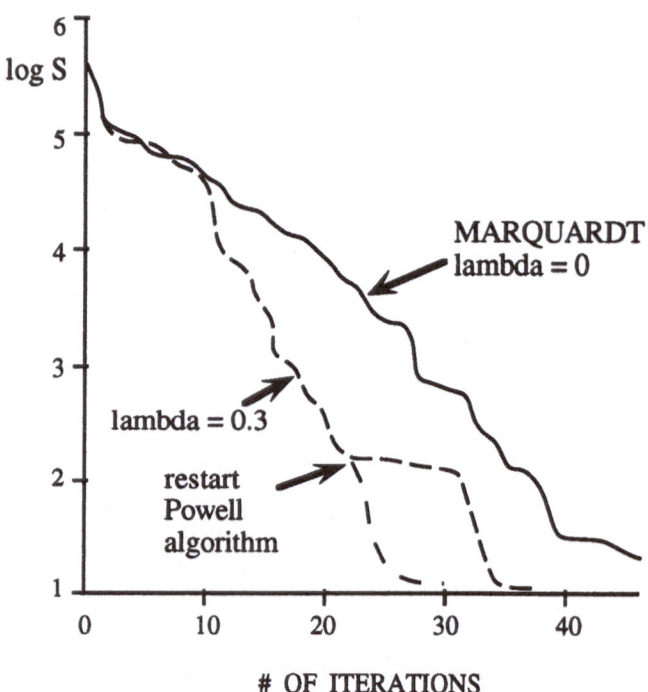

Fig. 4.30. Influence of parameter lambda on convergence in magnetic inversion using a multiplate model and the Marquardt optimization method. The variable optimization strategy with casual restart is also indicated. The objective function S is the sum of squares of error of fit at each data point. (McGrath and Hood 1973)

4.4.2 Gravity Inversion

Inversion of gravimetric data does not differ essentially from magnetic inversion, although the ambiguity problems call more often for additional constraining information. An early study on the use of optimization methods in gravity interpretation was undertaken by Al-Chalabi (1971). In the study a 2D polygonal model with additional regional background field was used (Fig. 4.31). $p_{poly} =$ $(x_{oi}, z_{oi}, \Delta\rho)$; i = 1, ..., 6 and $p_{add} = (\alpha = $ fixed, ZH).The starting values of the parameter may be taken from the "Smith rules" when no other information is available. Al-Chalabi found a variable optimization strategy with a direct search at the beginning and a gradient method (e.g. Davidon) at the later stage. In the example, two inversion results, which are given in Fig. 4.31, 12 parameters (co-ordinates of six corner points) of the polygon were obtained from 30 data values. The polygon is used to simulate a graben structure. The maximum number of iterations was limited to 150. The two models presented differ in their structure outside the graben area. In model (a) the graben material (less dense than the bedrock) extends towards both ends of the profile requiring a slightly higher regional background gravity field. In model (b) the graben is limited in extent horizontally, requiring a greater density contrast and a smaller regional gravity field.

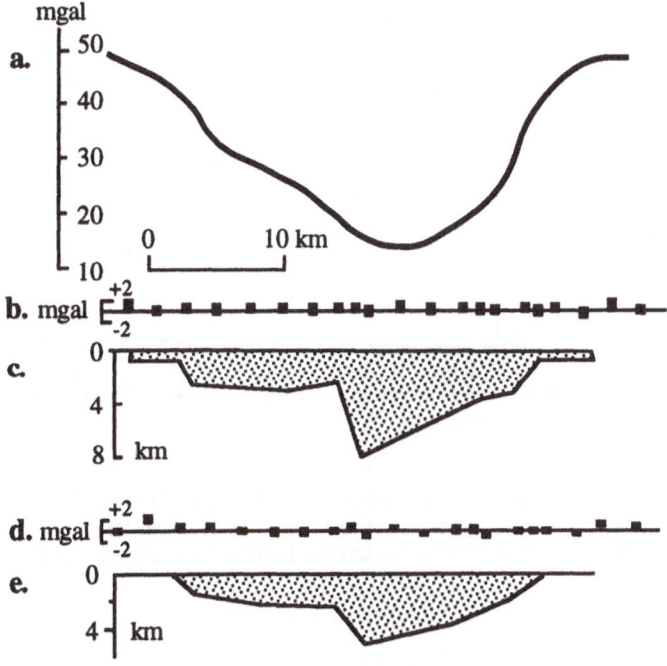

Fig. 4.31. Inversion of gravity data using a 2D polygonal graben model + constant regional field. Variable optimization strategy was adopted (Al-Chalabi 1971). **a. data** = observed anomaly profile; **b. squares** = residuals after fit of model 1; **c.** inverted model 1 (density contrast = - 0.25 g/cm^3, background = 51.0 mgal); **d.** residuals after fit of model 2; **e.** inverted model 2 (density contrast = - 0.27 g/cm^3, background = 50.0 mgal), where constant (zero) depth to the model surface outside the graben was fixed

4.4.3 Seismic Refraction

Seismic modelling has been concentrated very widely to use layered Earth models, especially until the tomographic approach became fashionable. One interesting exception has been an early paper by Ocola (1972a). The formulation was developed for a very general model, where the layer boundaries could have arbitrary topography. The parameter vector became rather involved, $\mathbf{p} = (x_{ij}, z_{ij}, \alpha_{ij}, \phi_j, v_j)$; $i = 1, 2, ..., N$ (# of seismic rays) and $j = 1, ..., M$ (the # of layers). The components of $\mathbf{p}$ can be deduced from simple geometrical considerations (cf. Ocola 1972a): x_{ij}, z_{ij}, differences of coordinates between subsequent points where each ray is crossing a layer boundary, ϕ_j the inclination of the layers at the same points and α_{ij} is the angle of arrival of the ith ray at the jth boundary, v_j the seismic velocities of the layers (Fig. 4.32). In addition, depths of the critical refraction point below each source and receiver were specified. The Marquardt algorithm and a least squares fit of the arrival times was used. The example is theoretical and uses a simple four-layer Earth model with two inclined boundaries (Fig. 4.33). The number of observations was 16 and 15 parameter values were deduced. The convergence reached the round-off error threshold in four iterations.

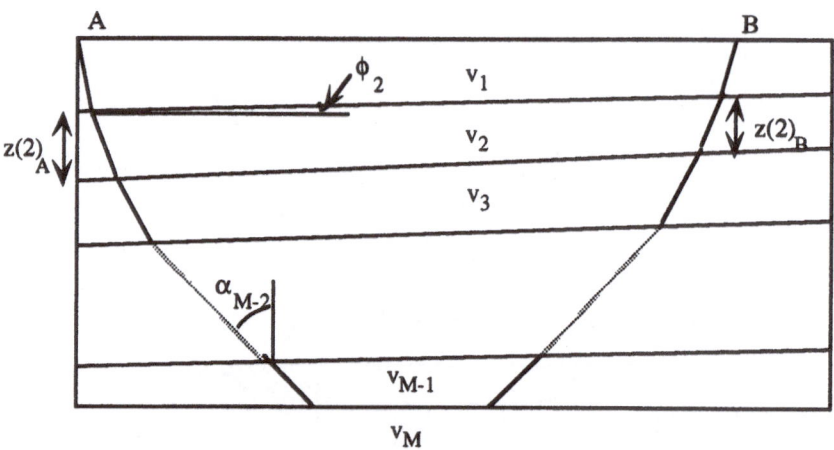

Fig. 4.32. Definition of the major model parameters used by Ocola (1972a) in non-linear inversion of refraction data. **A** Source (shot) point; **B** receiver point; if the refracted wave has changed type (from P to S, say) the left ray branch has different vertical angles than the right branch

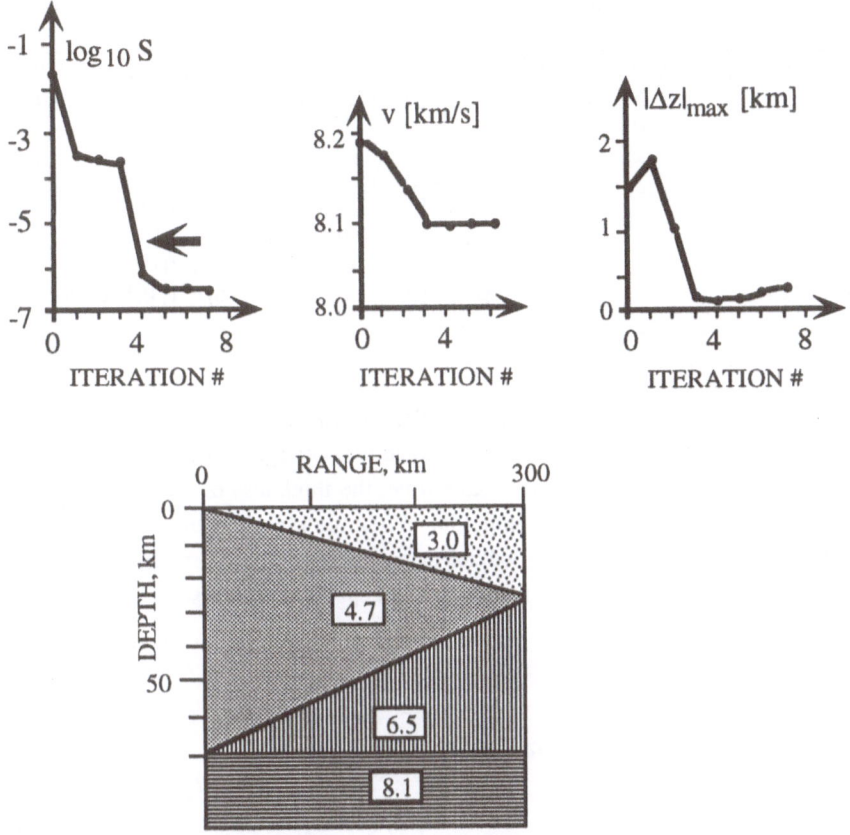

Fig. 4.33. Inversion of seismic refraction data using the non-linear Marquardt optimization technique. **Top:** Convergence of the logarithm of the objective function S **(left)**, of the bottom layer velocity **(center)** and the depth **(right)**. S is the sum of squares of the difference between model and measured arrival times. **Bottom:** The original model with velocities in km/s. (After Ocola 1972b)

4.4.4 Magnetotellurics

In a very early paper on computerized 1D inversion of magnetotelluric sounding data, Wu (1968) also applied the Marquardt technique. The layered Earth model parameters were the layer resistivities and thicknesses, thus $\mathbf{p_{MTS}} = (h_i, \rho_i)$; $i = 1, ..., N$. The "data" used in the inversion were theoretical apparent resistivities in the period range 0.01 to 10.000 s. The models labelled 'Wu model 1 ... 4' are explained in Table 4.6 and Figs. 4.34 - 4.39.

The convergence of the objective function S for the different models is shown in Fig. 4.34, the covergence of the layer resistivities in Figs. 4.35 - 4.37 and the layer thicknesses in Figs. 4.38 and 4.39. Rapid convergence occurred for model 1, whereas inversion of model 2 is problematic. By

TABLE 4.6   Models and parameters for magnetotelluric non-linear inversion (Wu 1968)

"Data" 1 and 3	$N = 3$	$T = 0.01 - 10\,000$ s	$p_{o1} = [(h_i = 1, 1, \text{fixed} = \infty); (\rho_i = 1, 0.1, 0.01)]$
2	$N = 3$	$T = 8.5 - 10\,000$ s	$p_{o1} = [(h_i = 1, 1, \text{fixed} = \infty); (\rho_i = 1, 0.1, 0.01)]$
4	$N = 3$	$T = 0.01 - 10\,000$ s	$p_{o3} = [(h_i = 1, 1, \text{fixed} = \infty); (\rho_i = 1, 0.01, 0.1)]$

Inverted models	1	$N = 3$	$p_{inv} = [(h_i = 1.1, 1.0, \infty); (\rho_i = 1, 0.1, 0.01)]$
	2	$N = 3$	$p_{inv} = [(h_i = 1.1, 0.1, \infty); (\rho_i = 1, 0.1, 0.01)]$
	3	$N = 4$	$p_{inv} = [(h_i = 0.986, 0.114, 1.0, \infty);$ $(\rho_i = \text{fixed} = 1, 1.1, 0.1, 0.01)]$
	4	$N = 3$	$p_{inv} = [(h_i = 1.0, 1.0, \infty); (\rho_i = 1, 0.1, \text{fixed} = 0.01)]$

leaving out the shorter periods below 8.5 s, the essential information content related to the resistivity of the uppermost layer is missing. Although ρ_1 ultimately converges towards it correct value, the convergence is slow and oscillatory. At the same time, the thickness of the intermediate layer 2 becomes underestimated by an order of magnitude. When the inversion model contains one extra layer (model 3), the result is astonishingly good, the sum of the two first layers approach the value 1.0 of the original model. The resistivity of the first layer was kept fixed, which soon brings the resistivity

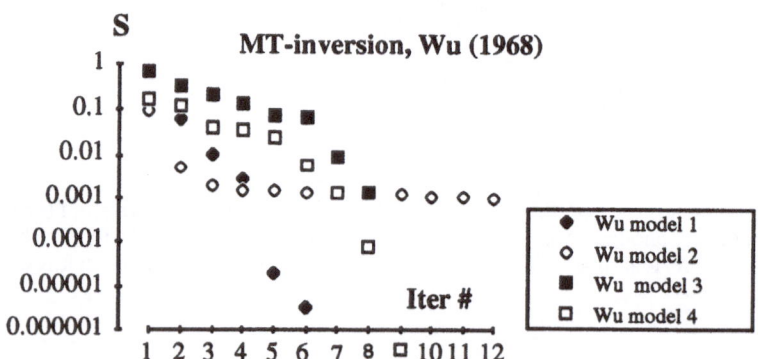

Fig. 4.34. Convergence of the objective function, S = sum of squares between "measured" and inverted apparent resistivities. The MTS models and their parameters are described in the text

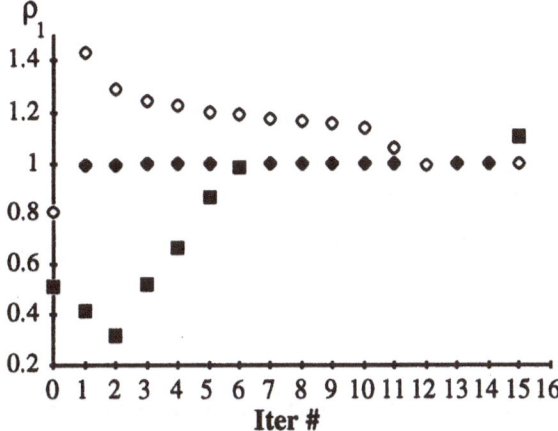

Fig. 4.35. Convergence of resistivity of the first layer. Model and inversion parameters are described in the text. For model 3, the resistivity of the "extra" second layer is shown. The **symbols** are the same as in Fig. 4.34 on the convergence of S

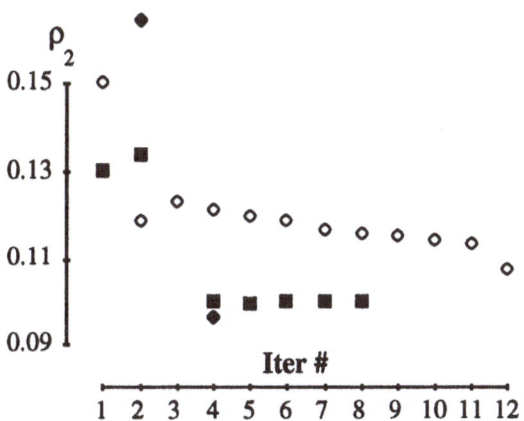

Fig. 4.36. Convergence of resistivity of the intermediate layer. Model and inversion parameters are described in the text. **Iteration # 1 = starting model. The symbols** are the same as in Fig. 4.34 on the convergence of S

of the "extra" layer to its "correct" value, $\rho_1 =1.0$. The other parameters approach their appropriate value at an early stage of iteration. The last model, 4, where the resistivities of the layers 2 and 3 are interchanged, does not converge properly, unless ρ_3 is kept fixed. The MTS "data" does not contain sufficient information about the more resistant layer below the conducting layer. The convergences of the various layer parameters have been collected into graphical form from the tables of parameter values given in the original paper by Wu (1968).

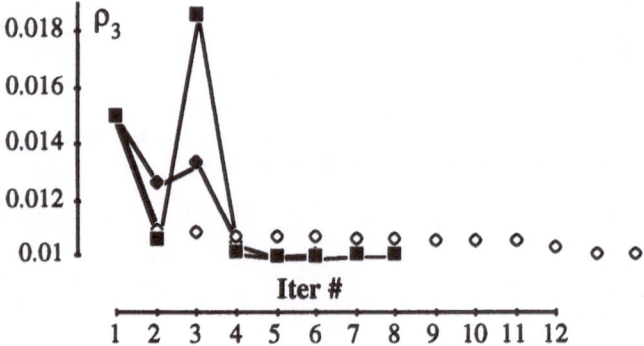

Fig. 4.37. Convergence of resistivity of the bottom layer. Model and inversion parameters are described in the text. **Iteration # 1 =** starting model. The **symbols** are the same as in Fig. 4.34 on the convergence of S

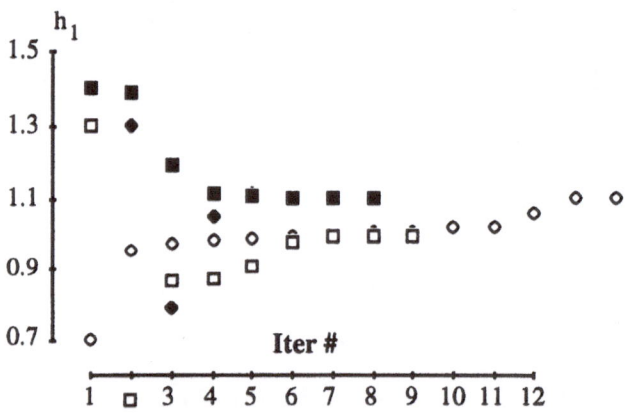

Fig. 4.38. Convergence of the thickness of the uppermost layer. For model 3, the sum of the first and the second "extra" layer has been shown. Model and inversion parameters are described in the text. **Iteration # 1 =** starting model. The **symbols** are the same as in Fig. 4.34 on the convergence of S

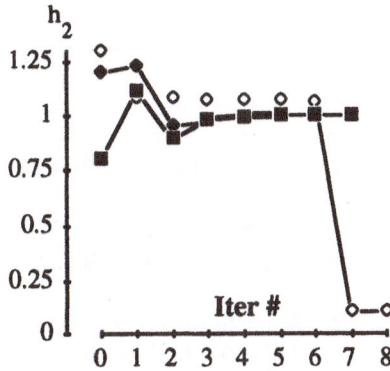

Fig. 4.39. Convergence of the thickness of the intermediate layer.
Model and inversion parameters are described in the text. **Iteration**
1 = starting model. The **symbols** are the same as in Fig. 4.34 on
the convergence of S.

REFERENCES

Aarts E, Korst J (1989) Simulated annealing and Boltzmann machines, a stochastic approach to combinatorial optimization and neural computing. John Wiley and Sons, New York

Albert A (1972) Regression and the Moore-Penrose pseudoinverse. Academic Press, New York, 180 pp

Al-Chalabi M (1970) Interpretation of two-dimensional magnetic profiles by non-linear optimisation. Bull Geof Teor Applic vol XII, 45-46: 3 - 20

Al-Chalabi M (1971) Some studies relating to nonuniqueness in gravity and magnetic inverse problems. Geophysics 36: 835 - 855

Al-Chalabi M (1972) Interpretation of gravity anomalies by non-linear optimization. Geophys Prosp 20: 1 - 16

Anderssen RS, Seneta E (1971) A simple statistical estimation procedure for Monte Carlo inversion in geophysics. PAGEOPH 91: 5 - 13

Anderssen RS, Seneta E (1972) A simple statistical estimation procedure for Monte Carlo inversion in geophysics. II: Efficiency and Hempel's paradox. PAGEOPH 96: 5 - 14

Becker K-H, Dörfler M (1989) Dynamical systems and fractals. Cambridge Univ Press, Cambridge, 398 pp

Bott MHP (1973) Inverse methods in the interpretation of magnetic and gravity anomalies. In: Bolt BA (ed): Methods in computational physics, vol 13, Academic Press, New York, pp 123 - 162

Conn AR (1982) Penalty function methods. In: Powell MJD (ed): Nonlinear optimization 1981. Proc of the NATO Advanced Research Institute, Cambridge, July 1981. Academic Press, Whitstable, Kent, 235 - 242

Cooper L, Steinberg D (1970) Introduction to methods of optimization. WB Saunders, Philadelphia, 381 pp

Daniels RW (1978) An introduction to numerical methods and optimization techniques. North-Holland, New York, 293 pp

Davidor Y (1990) An intuitive approach to genetic algorithms as adaptive optimization procedures. Techn Rep CS 90-07, The Weizmann Institute of Science, Rehovot, Israel, 25 pp

Draper NR, Smith H (1966) Applied regression analysis. John Wiley and Sons, USA, 407 pp

Fletcher R, Powell MJD (1963) A rapidly convergent descent method for minimization. Comput J 6: 163 - 168

Fletcher R, Reeves CM (1964) Function minimization by conjugate gradients. Comput J 7: 149 - 154

Fletcher R (1982) Methods for nonlinear constraints. In: Powell MJD (ed): Nonlinear optimization 1981. Proc of the NATO Advanced Research Institute, Cambridge, July 1981. Academic Press, Whitstable, Kent, 185 - 211.

Fletcher R (1987) Practical method of optimization. John Wiley and Sons, New York

Gay SP Jr (1963) Standard curves for interpretation of magnetic anomalies over long tabular bodies. Geophysics 28: 161 - 200

Gill PE, Murray W, Wright MH (1981) Practical optimization. Academic Press, London, 401 pp

Glover F (1989) Tabu search - part I. ORSA J Comput 1: 190 - 206

Glover F (1990) Tabu search - part II. ORSA J Comput 2: 4 - 32

Goldberg DE (1989) Genetic algorithms in search, optimization and machine learning. Addison-Wesley, Reading (MA)

Hillis WD (1987) The connection machine. Sci Am 256: 108 - 115

Hjelt S-E (1973) Experiences with automatic magnetic interpretation using the thick plate model. Geophys Prosp 21: 243 - 265

Hjelt S-E (1975) Performance comparison of non-linear optimization methods applied to interpretation in magnetic prospecting. Geophysica 13: 143 - 166

Hjelt S-E, Lanne E, Ruotsalainen A, Heiskanen V (1977) Regional interpretation of magnetic and gravimetric measurements based on combinations of dipping prisms and plates. Geol. Fören. Förh. 99: 216 - 225

Hooke R, Jeeves TA (1961) Direct search solution of numerical and statistical problem. J of Assoc Comput Machin 8:

Hopfield JJ (1985) Neural networks and physical systems with emergent collective computational abilities. Proc Natl Acad Sci USA. 79: 2554 - 2558

Jakobsen MO, Mosegaard K, Pedersen JM (1988) Global model optimization in reflection seismology by simulated annealing. In: Vogel A (ed) Model optimization in exploration geophysics 2. Proc 5th Int Mathematical Geophysics Seminar, Freie Univ Berlin, Feb 4 - 7, 1987. Fr Vieweg & Sohn, Braunschweig/ Wiesbaden, pp 361 - 381

Jones A (1970) Spiral - a new algorithm for non-linear parameter estimation using least squares. Comput J 13: 301 - 308

Jones AG, Hutton R (1979) A multi-station magnetotelluric study in southern Scotland- II. Monte-Carlo inversion of the data and its geophysical and tectonic implications. Geophys J R Astr Soc 56: 351 - 368

Keilis-Borok VJ, Yanovskaya TB (1967) Inverse problems of seismology (structural review). Geophys J R Astr Soc 13: 223 - 234

Kirkpatrick S, Gelatt SD Jr, Vecchi MP (1983) Optimization by simulated annealing. Science 220: 671 - 680

Kowalik J, Osborne MR (1968) Methods for unconstrained optimization problems. Elsevier, New York, 148 pp

Laarhoven PJM van, Aarts EHL (1987) Simulated annealing: theory and applications. D. Reidel, Dordrecht

Lakanen E (1975) An overall computer graphics software for geophysical profile interpretation. Paper presented at the 37th EAEG Meeting, Bergen, Norway.

Levenberg K (1944) A method for the solution of certain non-linear problems in least squares. Q Appl Math 2: 164 - 168

Levy AV, Montalvo A (1985) The tunneling algorithm for the global minimization of functions. SIAM J Sci Stat Comput 6: 15 - 29

Marquardt DW (1963) An algorithm for least-squares estimation of nonlinear parameters. J Soc Ind Appl Math 11: 431 - 441

McGrath PH, Hood PJ (1970) The dipping dike case: a computer curvematching method of magnetic interpretation. Geophysics 35: 831 - 848.

McGrath PH, Hood PJ (1973) An automatic least squares multimodel method for magnetic interpretation. Geophysics38: 349 - 358

Menke W (1989) Geophysical data analysis: Discrete inverse theory (revised ed). Academic Press, USA, 285 pp

Mora P (1988) Elastic wavefield inversion and the adjoint operation for the elastic wave equation. In: Vlaar NJ, Nolet G, Wortzel MJR, Cloetingh SAPL (eds): Mathematical geophysics. D. Reidel, Dordrecht, pp 117 - 137

Ocola LC (1972a) A non-linear least-squares method for seismic refraction mapping. Pt I. Algorithm and procedure. Geophysics 37: 260 - 272

Ocola LC (1972b) A non-linear least-squares method for seismic refraction mapping. Pt II. Model studies and performance of Reframap method. Geophysics 37: 273 - 287

Powell MJD (1981) Approximation theory and methods. Cambridge Univ Press, West Hanover, MA, 339 pp

Powell MJD (ed) (1982) Nonlinear optimization 1981. Proc of the NATO Advanced Research Institute, Cambridge, July 1981. Academic Press, Whitstable, Kent, 559 pp

Press F (1968) Earth models obtained by Monte-Carlo inversion. J Geophys Res 73: 5223 - 5234

Press F (1971) An introduction to Earth structure and seismotectonics. In: Coulomb J, Maputo M (eds): Proc. of the Int School of Physics Enrico Fermi, Course L, Mantle and core in planetary physics. Academic Press, New York, 209 - 241

Rothman DH (1985) Nonlinear inversion, statistical mechanics and residual statics estimation. Geophysics 50: 2797 - 2807

Rothman DH (1986) Automatic estimation of large residual statics corrections. Geophysics 51: 332 - 346

Tarantola A (1987) Inversion of travel times and seismic waveforms. In: Nolet G (ed.) Seismic tomography. (With applications in global seismology and exploration geophysics). D. Reidel, The Netherlands, 135 - 157

Weisberg S (1985) Applied linear regression.John Wiley and Sons, New York, 324 pp

Whitehill DE (1973) Automated interpretation of magnetic anomalies using the vertical prism model. Geophysics 38: 1070 - 1087.

Whittall KP (1986) Inversion of magnetotelluric data using localized conductivity constraints. Geophysics 51: 1603 - 1607

Wiggins RA (1969) Monte-Carlo inversion of body-wave observations. J Geophys Res 74: 3171 - 3181

Wilde DJ (1964) Optimum seeking methods. Prentice-Hall, Englewood Cliffs, NJ, 202 pp

Wilde DJ, Beightler CS (1967) Foundations of optimization. Prentice-Hall, Englewood Cliffs, NJ, 480 pp

Wu FT (1968) The inverse problem of magnetotelluric sounding. Geophysics 33: 972-9

Wunsch C (1978) The North Atlantic general circulation west of 50° W determined by inverse methods. Rev Geophys Space Phys 16: 583 - 620

Chapter 5

MAXIMUM LIKELIHOOD
and
MAXIMUM ENTROPY

CHAPTER 5

MAXIMUM LIKELIHOOD and MAXIMUM ENTROPY

5.1   INTRODUCTION AND DEFINITIONS

The classical correlation method appeared in some early papers when geophysical inversion using computers started (see Beskow and Granar 1969; Naudy 1971; Olszak and Peschel 1971) and also the autoregressive properties of anomaly profiles has been studied (Beryland and Roze 1971). Statistical principles of geophysical interpretation have been the subject of prof. Holzmann and his group already since late 1960s (see Kalinina and Holzmann 1967; Holzmann and Protopopova, 1968; Kalinina 1970; Holzmann 1971; Holzmann, 1976). They applied the technique both in the inversion of seismological as well as potential field (magnetic and gravity) data. Two textbooks on the subject (Holzmann 1982; Holzmann and Kalinina 1983) do not seem to have penetrated the language barrier of English-dominating geophysical scientific journals. A renaissance of statistical considerations in geophysical inversion started with two remarkable papers by Tarantola and Valette (1982a, b) and the book by Tarantola (1987).

General considerations of the statistical approach can be found in Rao (1965) and geophysical formulations in Peschel (1971, 1972). Information theory and the maximum entropy method are discussed in Burg (1972), Deeming (1987), Goldman (1955), Jackson and Matsu'ura (1985) and Smith and Grandy (1985).

5.1.1   Probability Density

Let us consider the normalized one-dimensional probability density function $P(x)$ as defined by Tarantola (1987; cf. also Menke 1989):

$$s_p(m) = \left[\int_{-\infty}^{\infty} \left(P(x) \, |x - m|^p \, dx \right) \right]^{1/p} \tag{5.1}$$

This can be used to define characteristic statistical estimators of the distribution. The value of m, which minimizes $s_p(m)$ is called the *centre* of $P(x)$ and the corresponding minimum value of $s_p(m)$ the *dispersion* of $P(x)$. Depending on the chosen norm (or the value of p), one has (Tarantola 1987):

$p = 1$ m_1 = median $s_1(m)$ = mean deviation

$p = 2$ m_2 = mean $s_2^2(m)$ = standard deviation

$p = \infty$ m_∞ = mid-range $s_\infty(m)$ = half-range

In more detail:

$$1/2 = \int_{-\infty}^{m_1} P(x)\, dx = \int_{m_1}^{\infty} P(x)\, dx; \quad s_1 = \int_{m_1}^{\infty} x\, P(x)\, dx - \int_{-\infty}^{m_1} x\, P(x)\, dx$$

$$m_2 = \int_{-\infty}^{\infty} x P(x)\, dx$$

$$(5.2)$$

$$s_2^2 = \int_{-\infty}^{\infty} x^2\, P(x) dx - m_2^2$$

$$m_\infty = (x_{Max} + x_{min})/2; \qquad\qquad s_\infty = (x_{Max} - x_{min})/2.$$

The generalization to the multidimensional case for $p = 2$ involves the covariance operator:

$$C_2^{ij}(m) = \int_{-\infty}^{\infty} (x^i - m^i)(x^j - m^j)P(x)\; dx. \qquad\qquad (5.3)$$

The vector m_2 minimizing the diagonal elements of $\underline{C}_2$ are the mean of x and the diagonal elements themselves the variances (equalling the square of standard deviations). The quantity:

$$\| x \|_2 = (x^T \underline{C}_2^{-1} x)^{1/2}, \qquad\qquad (5.4)$$

is the weighted l_2 norm of the vector x. Normalizing $\underline{C}_2$ produces the correlation matrix $\mathbf{R}$ with the elements:

$$\rho_2^{ij} = \frac{C_2^{ij}}{\sigma_2^i \sigma_2^j}. \qquad\qquad (5.5)$$

The normalized probability density function with minimum information content is:

$$F_p(x) = \frac{p^{1-1/p}}{2\sigma_p\, \Gamma(1/p)}\, \exp\left(-\frac{1}{p}\frac{|x - m_p|^p}{(\sigma_p)^p}\right). \qquad\qquad (5.6)$$

With $p = 2$ F_p becomes the normal distribution, we have the Gaussian function F_2. For $p = \infty$ the probability density function is a box-car of width $2\sigma_p$ and for $p = 1$ a double exponential centred around m_1.

Given a normalized probability density function P(x) and a reference distribution $P_N(x)$, the relative information content of P(x) (with respect to $P_N(x)$) is:

$$I(P, P_N) = \int P(x) \log_b[P(x)/P_N(x)] \, dx. \qquad (5.7)$$

Depending on the base b used for the logarithm, the unit of information content I is either a *bit* (b = 2), a *nep* (b = e = 2.71828...), or a *digit* (b = 10). If the null distribution has been chosen as reference I is called simply the information content of P(x). This definition is a straightforward extension of the original definition of Shannon for discrete probabilities.

The information content has the following properties:

1. Information of the reference (null) distribution is zero.
2. I > 0 for all other distributions.
3. The sharper peaked P is, the greater I is.
5. I is invariant under reparametrization.

The amount of information added to an inverse problem can therefore be measured by imposing a priori constraints.

5.1.2 Measure of Information

Let us reconsider the concept of transmitted information from the classical point of view of Shannon (1948). The amount of transferred information I can be defined using probabilities:

$$I = \log \left(\frac{P_{vo}}{P_o} \right), \qquad (5.8)$$

where P_o is the probability at the reception point before the arrival of a message and P_{vo} the probability of at the reception point after the arrival of a message. If the transmission is error-free, $P_{vo} = 1$ and $I = -\log(P_o)$. As a simple example, consider a family waiting for a message of the birth of their first grandchild. Besides the health of the child and mother, one piece of information everybody wants to have as quick as possible is the sex of the newborn. The probability of the birth of a boy and a girl is about equal (at least with accuracy enough for this simple example), $P_o(boy) = 0.5$ and $P_o(girl) = 0.5$. When the child is born and the message reaches the grandparents, the amount of information received about the sex of the child is $I = -\log(0.5) = \log 2$.

Let us consider another situation of more geophysical character. We are looking at rock samples taking from boreholes of a research area. The samples have been weighted and classified according to their weight into classes (Fig. 5.1).

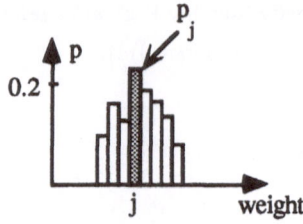

Fig. 5.1. Borehole samples are grouped according to their weight into classes **j**. The percentage of samples within each class, **p(j)**, approaches the probability distribution of the samples provided the number of samples increases sufficiently

The probability of a new sample to belong to any of the weight classes is naturally = 1 (provided of course the choice of classes covers the samples of the area with reasonable completeness). Therefore, the amount of information obtained when a new sample is weighted and found to belong to class j, is:

$$I = - \log p_j .$$

Suppose we have received m new samples, then the expected number of samples belonging to the class j is $N_j = m \, p_j$ and the amount of information for these N_j occurrences is:

$$I_j = N_j \, (- \log p_j) = - m \, p_j \log p_j.$$

The total information obtained after having weighted and classified the m samples is

$$I_{tot} = - m \sum_j p_j \log p_j. \qquad (5.9)$$

The average information or information per number of samples is clearly:

$$H = I_{tot}/m = - \sum_j p_j \log p_j. \qquad (5.10 \, a)$$

The quantity H is called, in analogy to its mechanistical counterpart in statistical physics, *entropy*. This is the original definition of information given by Shannon. The entropy is a property which always tends to increase.

If the channel carrying the information is not noise-free, we will need to take into account additional probabilities, although the basic definition of information and entropy remains. We denote the probability that a message i is transmitted with P_i, the probability of receiving a message j with P_j and finally the probability of receiving message j, when the message i is transmitted with P_{ij}. We then have:

$$P_j = \sum_i P_i \, P_{ij}, \qquad \text{(summation over all transmitted messages)}$$

and $\sum_j P_j = 1$. (probability of receiving any of the messages is 1).

The amount of information is:

$$I = \log \left(\frac{P_i P_{ij}}{P_j} / P_i \right) = \log (P_{ij}) - \log (P_j),$$

and the entropy

$$H = \sum_j \sum_i P_i P_{ij} \, I = H_j - H_{ij}, \qquad (5.10\,b)$$

where H_j is the entropy, when the transmitted message is unknown and H_{ij} the entropy, when the message is known.

For a continuous signal, the summation in the entropy becomes an integration over all possible states

$$H = - \int P(x) \log[P(x)] \, dx. \qquad (5.10\,c)$$

In the *maximum entropy method* (MEM) of geophysical inversion (or any other inversion for that matter), x is replaced by the unknown parameter p and the probability is its (assumed) statistical distribution, which will be associated with some constraints. The entropy takes the place of the objective function (cf. non-linear optimization methods).

5.2   PROBABILISTIC FORMULATIONS OF THE INVERSION PROBLEM

5.2.1   The Principle of Maximum Likelihood

In the preceding chapters, the model parameters were considered as separate entities, components of a discrete, finite vector. Higher level principles of interpretation require another approach. Not only the data, but also the parameters are considered to have a statistical distribution. Before going into a detailed survey of statistical methods of inversion, it is instructive to consider the whole problem in a general framework of probability theory. The description follows the lines set out by Peschel (1971).

Let us define four functions (or operators, if you wish):

F_h : the data, which may consist of subgroups F_{hv} one group for each (eg. each F_{hv} is a data set taken by a different geophysical method).

U: a complete description of the subsurface of the research area.

M_i: a model of the subsurface of the research area. Each model M_i (i = 1, 2, ... K) can be considered as a different realization (approximation) of U.

F_t: the calculated (theoretical) geophysical field related to a model M of the subsurface. Subgroups F_{tv} correspond to the various geophysical methods as for the data.

The geophysical interpretation consists of finding a model sufficiently close to the "real truth", the complete description U of the subsurface. In the ideal case M ≡ U. Let us define the following probabilities:

$P\{M_i\}$: the probability of the description of a model M_i before data is available
$P\{F_h|M_i\}$: the conditional probability that the model M_i fits the data $[F_t(M_i) = F_h]$
$P\{M_i|F_h\}$: the descriptive probability of the model M_i after the data have become available.

Alternatively, the second probability can be written $P\{F_h|F_t\}$. $P\{M_i\}$ is the *a priori* probability and $P\{M_i|F_h\}\}$ the *a posteriori* probability.

The principle of *maximum likelihood* (ML) requires us to choose that model, which maximizes the a posteriori probability. Using the equations of conditional probabilities of statistics (Bayes theorem) the principle is stated as:

$$P\{M_i|F_h\} = \frac{P\{M_i\}\cdot P\{F_h|M_i\}}{\sum_j P\{M_{ij}\}\cdot P\{F_h|M_{ij}\}} = \max!$$
(5.11)

Here, M_{ij} is the jth hypothesis (or realization) of the ith model. If several models of data sets from several methods are involved in the best interpretation, then the probabilities have to be multiplied

$$\prod_i P\{M_i|F_h\} = \max! \qquad \text{or} \qquad \prod_v^K P\{M_i|F_{hv}\} = \max!$$

Normally only a single hypothesis about the model is made: that the model description of the subsurface is true. The counter hypothesis is exclusive and accordingly its probability is $P\{M_i \neq U\} = 1 - P\{M_i = U\}$. $P\{M_i \neq U\}$ enters the denominator of the probability to be maximized:

$$P\{M_i|F_h\} = \frac{P\{M_i\}\cdot P\{F_h|F_t\}}{P\{M_i\}\cdot P\{F_h|M_i\} + [1 - P\{M_i\}]\cdot P\{F_h|\bar{\bar{F}}_t\}} .$$
(5.12)

Let us now assume that the data contain uncorrelated Gaussian noise:

$$F_h = F_t(M_i) + F_e .$$

$$F_e = N(o, \sigma^2).$$

(5.13)

Then:

$$P\{F_h|F_t\} = A\, e^{-q\sum\limits_{\mu}^{N} F_{t\mu}^2} \qquad \text{with } A = \frac{1}{\sqrt{2\pi\sigma^2}}; q = \frac{1}{2\sigma^2}$$

(5.14)

and

$$P\{F_h|\bar{F}_t\} = A\, e^{-q\sum\limits_{\mu}^{N} F_{h\mu}^2} .$$

If the data errors correlated with each other, the probability density function would contain the correlation matrix **R** and then:

$$A \rightarrow A = \frac{R^{-1}}{2}; q = \frac{|R|^{1/2}}{(2\pi)^N} .$$

(5.15)

By introducing

$$Q = \frac{P\{F_h|F_t\}}{P\{F_h|\bar{F}_t\}} = e^{-q\sum\limits_{\mu}^{N} F_{t\mu}^2} \cdot e^{2q\sum\limits_{\mu}^{N} F_{t\mu}F_{h\mu}},$$

the probability to be maximized, (5.11) can be written as:

$$P\{M_i|F_h\} = \frac{P\{M_i\}\cdot Q}{P\{M_i\}\cdot Q + 1 - P\{M_i\}} = \frac{1}{\frac{1}{Q}(\frac{1}{P\{M_i\}} - 1) + 1} = max!$$

By examining the various expressions for $P\{M_i|F_h\}$, the principle of maximum likelihood is realized if any of the four following conditions are fulfilled:

1. $P\{M_i|F_h\} = max!$

2. $Q = max!$

3. $\sum\limits_{\mu} F_{h\mu}F_{t\mu} = max!$

4. $P\{F_h|M_i\} = min!$

The third condition tells that the **correlation** between the data and the model field sets has to be a maximum. The last condition again is equal to $\sum_{\mu} F_{e\mu}^2 = $ min!, the familiar requirement of minimizing the sum of squares of the data errors. Thus a quite general exposition of the principle of maximum likelihood (ML) says that the classical method of least squares produces the *statistically most probable solution*, when the data errors (model fitting error) are uncorrelated and normally distributed. Thus the popular LSQ method has a firm and solid background in statistical theory of estimation. Another question is, how often the requirement of Gaussian errors of data is true for real geophysical fields. The probabilistic approach to the inversion problem is then needed and the newest development of Bayesian estimation can provide some answers also for other probability functions (cf. references).

The formalism allows also to consider the comparison of two models M_1 and M_2 statistically with each other (Olszak and Peschel 1971). Then $F_{e\mu}$ is replaced by $F_{ei\mu} = (F_h - F_{ti})\mu$ [i = 1, 2] and $P\{F_h|\overline{F}_t\}$ by:

$$P\{F_h|M_2\} = A\, e^{-q\sum_{\mu}^{N} F_{e2\mu}^2}.$$

$$(5.16)$$

The factor $Q = e^{-q\sum_{\mu}^{N} (F_{t1\mu}^2 - F_{t2\mu}^2)} \cdot e^{2q\sum_{\mu}^{N} [F_{h\mu}\cdot(F_{t2\mu}-F_{t1\mu})]}$ and either of the two extrema will fulfill the principle of maximum likelihood:

$$\sum_{\mu}^{N} [F_{h\mu}\cdot(F_{t2\mu}-F_{t1\mu})] = \text{max!}$$

or $$(5.17)$$

$$\sum_{\mu}^{N} (F_{e1\mu}^2 - F_{e2\mu}^2) = \sum_{\mu}^{N} [(F_{h\mu}-F_{t1\mu})^2 - (F_{h\mu}-F_{t2\mu})^2] = \text{min!}$$

5.2.2   The Maximum Entropy Method

Let us assume that a geophysical model parameter obeys a (assumed or known) statistical distribution. The entropy becomes an integration over all possible states

$$H = -\int P(p)\, \log[P(p)]\, dp,$$

$$(5.18)$$

and the statistical distribution of p will be associated with some (well-known) constraints. In the maximum entropy method (MEM) of geophysical inversion (or any other inversion for that matter), the entropy takes the place of the objective function (cf. non-linear optimization methods). MEM can be defined as:

$$S = H(P, p) = - \int_a^b P(p) \log [P(p)] \, dp = \text{max!} \qquad (5.19)$$

with constraints

$$K_i = \int_a^b \Phi_i(P, p) \, dp. \qquad (5.20)$$

The MEM method has received its name from statistical mechanics.

The definitions above are easily generalized to the case of several parameters. Following Menke (1989) let us consider the underdetermined linear problem $\underline{G} \, p = F$, where $P(p)$ is taken as unknown. Furthermore assume, that the data are exactly equal to the mean (expected) value of the model parameters, i.e. $\underline{G} \, E(p) = F$. Then MEM can be formulated as follows [with $P_N(p)$ being the null distribution = the state of complete ignorance]:

Find the $P(p)$, that maximizes $I[P(p), P_N(p)]$ subject to the constraints:

$$\underline{G} \, E(p) = F$$

$$\int P(p) \, \partial p = 1.$$

For example for $\Phi_i = P(p)$, then $K_i = 1$. The method of Lagrangian multipliers can be used to transform the constrained optimization problem into a single (non-linear) equation

$$\frac{\partial H}{\partial P} + \sum_i \lambda_i \frac{\partial \Phi_i}{\partial P} = 0. \qquad (5.21)$$

Menke (1989) also points out that it seems that MEM does not require knowledge of the a priori distribution. However, nonuniqueness of the definition of information makes MEM equally arbitrary as other methods.

Let us consider a simple example (assuming for simplicity that again p is a scalar). The data contain an error, which is Gaussian with a variance σ and the range of the parameter p runs from $- \infty$ to $+ \infty$. We have:

$$H(P, p) = - \int_{-\infty}^{+\infty} P(p) \log [P(p)] \, dp = \text{max!} \qquad (5.22)$$

$$1 = \int_{-\infty}^{+\infty} P(p) \, dp, \qquad (5.23 \, a)$$

$$\sigma^2 = \int_{-\infty}^{+\infty} p^2\, P(p)\, dp. \tag{5.23 b}$$

Obviously, $\partial H/\partial P = -(1 + \log P)$, $K_1 = 1$, $K_2 = \sigma^2$, $\Phi_1 = P(p)$, $\Phi_2 = p^2\, P(p)$, $\partial\Phi_1/\partial P = 1$ and $\partial\Phi_2/\partial P = p^2$. The Lagrangian equation becomes:

$$-(1 + \ln P) + \lambda_1 + \lambda_2\, p^2 = 0,$$

which can be solved for P as a function of the two other unknowns, the Lagrangian multipliers. The expression:

$$P(p) = e^{\lambda_1 - 1}\cdot e^{\lambda_2 p^2}$$

is substituted into the first constraining Eq. (5.23 a). The first exponential function does not depend on the integration variable and the remaining integral can be solved in closed form.

$$1 = \int_{-\infty}^{+\infty} e^{\lambda_1-1}\cdot e^{\lambda_2 p^2}\, dp = 2\, e^{\lambda_1-1}\cdot\int_{-\infty}^{+\infty} e^{\lambda_2 p^2}\, dp = 2\, e^{\lambda_1-1}\cdot\frac{1}{2}\sqrt{\frac{\pi}{-\lambda_2}}\ .$$

The result gives an equation between the two Lagrangian multipliers, but it is sufficient to solve for the exponential function only, since the second constraint Eq (5.23 b) depends on the first parameter only through this function. Substitution and integration give:

$$\sigma^2 = \int_{-\infty}^{+\infty} p^2\sqrt{\frac{-\lambda_2}{\pi}}\cdot e^{\lambda_2 p^2}\, dp = 2\sqrt{\frac{-\lambda_2}{\pi}}\cdot\int_{-\infty}^{+\infty} p^2 e^{\lambda_2 p^2}\, dp = 2\sqrt{\frac{-\lambda_2}{\pi}}\cdot\frac{1}{4}\sqrt{\frac{\pi}{(-\lambda_2)^3}} = -\frac{1}{2\lambda_2}.$$

Collecting the pieces together we have:

$$P(p) = \frac{1}{\sqrt{2\pi\sigma^2}}e^{-\frac{p^2}{2\sigma^2}}$$

showing that for normally distributed data errors the maximum entropy method (MEM) and the principle of maximum likelihood (ML) are equivalent.

5.2.3 Bayesian Estimation

The Bayesian approach can be considered as a modification (generalization) of the least-squares method, where a priori information about the parameters enters into the optimization scheme (Deeming 1987). It is probably more correct to state the other way round, that the least-squares approach is a special case of Bayesian estimation.

The basic Bayes' theorem concept is, that for a set of data, in our case the data vector $\mathbf{F}_h$, there exists a vector of the unknown parameters, $\mathbf{p}$, the conditional probability of the parameters, which, given the data, $P(\mathbf{p}|\mathbf{F}_h)$, can be written as:

$$P(\mathbf{p}|\mathbf{F}_h) = P(\mathbf{p})\ P(\mathbf{F}_h|\mathbf{p})\ /\ P(\mathbf{F}_h). \qquad (5.24\ a)$$

The denominator is a normalizer, which is independent of the choice (optimization) of the parameters. Therefore it is sufficient to consider only a part of the above equation:

$$P(\mathbf{p}|\mathbf{F}_h) \approx P(\mathbf{p})\ P(\mathbf{F}_h|\mathbf{p}) \qquad (5.24\ b)$$

The a priori probability, $P(\mathbf{p})$, may not necessarily be known, in which case it has to be assumed. A constant, independent of $\mathbf{p}$, is a commonly used choice, which leads to the maximum likelihood method when the a posteriori probability $P(\mathbf{p}|\mathbf{F}_h)$ is maximized with respect to $\mathbf{p}$. In Bayesian estimation some other choice, based on a priori knowledge or some basic idea about how the parameters might look, is made.

The probability function $P(\mathbf{F}_h|\mathbf{p})$ is called the likelihood function and we need a statistical model in order to compute it. Using the traditional choice of uncorrelated, multivariate Gaussian distribution with constant dispersion σ, we have:

$$P(\mathbf{F}_h|\mathbf{p}) = \prod_i^N e^{-\ e_i^2/2\sigma^2} = e^{-\ S(\mathbf{p})/2\sigma^2} \qquad (5.25)$$

Without going into further details, let us finally discuss some useful expressions for the probabilistic approach of geophysical inversion. For additional details see the recent literature, some of which can be found in the references of Chapter 8.

5.2.4   Some Useful Expressions for the Probabilistic Approach of Inversion.

5.2.4.1   Normal Distribution

Following mainly the exposition of Menke (1989), let us start with the classical normal distribution. When all parameters are normally distributed, the a priori density function is

$$\text{ß}(x) = \exp\{ - [(x-x_0)^T\ \underline{C}_0^{-1}\ (x-x_0)]/2\} \qquad (5.26\ a)$$

where x_0 is the a priori expected value and $\underline{C}_0$ the a priori covariance matrix. Replacing x with the parameter vector one has:

$$p(x) = \exp - [(x-x_0)^T\ \underline{C}_0^{-1}\ (x-x_0)]/2 \qquad (5.26\ b)$$

For a linear inversion problem $F \cdot x = 0$, whence:

$$T(x) = \exp \{ - [(F \cdot x)^T \underline{C}^{t-1} (F \cdot x)]/2 \} \tag{5.27}$$

$$\mu(x) = \text{const.}$$

$$\beta(x) = \exp \{ - [(x-x_0)^T \underline{C}_0^{-1} (x-x_0) + (F \cdot x)^T \underline{C}^{t-1} (F \cdot x)]/2 \} \tag{5.28 a}$$

Furthermore:

$$\beta(x) = \exp \{ - (x-x^*)^T \underline{C}^{*-1} (x-x^*) \} \tag{5.28 b}$$

with
$$x^* = P \cdot x_0$$
$$\underline{C}^* = P \cdot \underline{C}_0$$

and
$$P = 1-Q$$
$$Q = \underline{C}_0 F^{-1}(F \underline{C}_0 F^T + \underline{C}_t)^{-1} \cdot F \ .$$

Thus, the a posteriori density function is also normally distributed and its expected value is x^* and the covariance matrix $\underline{C}^*$. If theoretical errors can be neglected, then $\underline{C}_t = 0$. Alternatively written:

$$x = [d,p]^T \qquad\qquad x_0 = [d_0, p_0]^T$$

$$x^* = [d^*, p^*]^T \tag{5.29}$$

$$\underline{C}_0 = \begin{bmatrix} \underline{C}_{dd} & \underline{C}_{dp} \\ \underline{C}_{pd} & \underline{C}_{pp} \end{bmatrix}$$

$$F \cdot x = [\underline{I} - \underline{G}] [d,p]^T = d - \underline{G} \cdot p = 0 \ ,$$

which is the notation usually found in the literature.

5.2.4.2 Smoothness

The "smoothness" l of the parameter vector p can be defined using the multiplying matrix $\underline{D}$:

$$l = \underline{D} \cdot p \tag{5.30}$$

where, e.g.

$$D = [\begin{matrix} -1 & 1 & .... \\ .. & -1 & 1 & .... \\ ...... & .... \\ .... & -1 & 1 \end{matrix} \qquad]$$

The "smoothness" of the solution is

$$L = l^T \cdot l = (D \cdot p)^T (D \cdot p) = p^T \cdot D^T \cdot D \cdot p = p^T \cdot W_m \cdot p \qquad (5.31)$$

5.2.4.3 Weighted Least Squares

If the accuracy of the measurements varies from point to point a weighted least squares can be defined by:

$$S = \varepsilon^T \cdot W_e \cdot \varepsilon \qquad (5.32)$$

$$S = min!$$

$$-> \qquad p^* = [G^T \cdot W_e \cdot G]^{-1} \cdot G^T \cdot W_e \cdot F_h \qquad [F_h = G \cdot p]$$

The weighted length of the fit is obtained in the form:

$$L = [p - <p>]^T \cdot W_m \cdot [p - <p>] \qquad (5.33)$$

$$p^* = <p> + W_m \cdot G^T \cdot [G \cdot W_m \cdot G^T]^{-1} [F_h - G \cdot <p>]. \qquad (5.34)$$

5.2.4.4 Parameter Boundaries

Let the matrix F describe the boundaries of the parameter values. The of the matrix has a value of 1 for those parameters, which are given prespecified values, which are kept fixed during the inversion.

$$F \cdot p = h \qquad (5.35)$$

$$F = [1 \ 1 \ 1 \ .... \ 1]/M \qquad \text{all parameters specified in advance}$$

$$F = [0 \ 1 \ 0 \ 0 \ .....0] \qquad \text{second parameter specified in advance}$$

etc.

Using the Lagrangian multipliers:

$$S(\mathbf{p}) = \epsilon^T \cdot \epsilon + 2 \sum_i^p [\lambda_i \sum_j^M (F_{ij} p_j - h_i)] \qquad (5.36)$$

$$\partial S / \partial p_q = 0 \qquad (5.37)$$

$$\begin{bmatrix} G^T \cdot G & F^T \\ F & 0 \end{bmatrix} [\mathbf{p}] = [\begin{matrix} G^T \cdot y_h \\ h \end{matrix}]$$

5.2.4.5   More on A Priori Information

It may be useful to compare the development presented by Tarantola and Valette (1982a). They enlarged the original information concept of Shannon in the following way. Given the probability density function [p.d.f] $f(x)$, the amount of information is:

$$I(f;\mu) = \int f(x) \, \text{Log}\{f(x)/\mu(x)\} \, dx , \qquad (5.38)$$

where μ is the **pdf** of complete ignorance **ttf** (Note! this definition is not included in the original definition of information by Shannon).

The definition (5.38) has following properties:

1. $I(f; \mu) = I(f'; \mu')$ -> I is invariant with respect to a change of variables
   (Not used by Shannon!).

2. $I(f; \mu) \geq 0$ -> Information cannot be negative.

3. $I(\mu; \mu) = 0.$ -> In the case of complete ignorance the information is 0.

   Also $I(f; \mu) = 0$ --> $f = \mu$ holds true.

If s_i and s_j are two states of information, independent of each other and with the distribution functions f_i and f_j respectively and μ corresponds to the state of zero information, the combined state of information is then the conjunction distribution $f(x)$:

$$f(x) = f_i(x) \, f_j(x) / \mu(x) . \qquad (5.39)$$

Define $\mathbf{x}$ so that it consists of two parts, the observations $\mathbf{d}$ and the model parameters $\mathbf{p}$. A pdf $r(\mathbf{d}, \mathbf{p})$, called the *a priori* density function, is associated with each of these sets. The observations and the model parameters are usually connected via a functional dependence (the equation for the theoretical anomaly) or:

$$F(x) = F(d,p) = 0 \qquad\qquad \text{or } d = G(p) \qquad\qquad (5.40)$$

If the model is not known exactly, then instead of Eq. (5.40) one has to define the conditional pdf of d as:

$$T(x) = T(d \mid p).\mu(p) . \qquad\qquad (5.41)$$

In the case of exact fit $T(d \mid p) = \delta[d - G(p)]$. If the function $T(d \mid p)$ cannot be given analytically, the choice:

$$T (d \mid p) = \text{const} . \exp\{ - \|d-G(p)\|^2/(2\sigma^2)\} \qquad\qquad (5.42)$$

is made. The conjunction of this new pdf and r defines the a posteriori pdf.

$$\beta(x) = r(x)\, T(x)\, /\, \mu(x) \qquad\qquad (5.43)$$

Since the a priori information on the data is usually independent of the a priori information on the parameters, one is able to write:

$$r(d,p) = \Sigma d(d)\, \Sigma p(p) \qquad\qquad (5.44)$$

Combining Eq. (5.43) and Eq. (5.44) and by integrating, one has:

$$\beta p(p) = \Sigma p(p) \ddot{}\{ \Sigma d(d)\, T(d \mid p)/ \mu d (x)\}\, dd = \qquad\qquad (5.45)$$

$$= \Sigma p(p)\, \Sigma d(G(p)) / \mu d (G(p)) \qquad (\text{exact fit}) . \qquad\qquad (5.46)$$

The a posteriori expectation value is:

$$E(P) = \int p\, \beta p(p)\, dp , \qquad\qquad (5.47)$$

or the a posteriori covariance matrix:

$$C = E\{ (P-E(P)).(P-E(P))T\} = \int p\, pT\beta p(p)\, dp - E(P)\, E(P)T. \qquad\qquad (5.48.)$$

Examples on the use of and the importance of a priori information are discussed further in Chapter 8.

REFERENCES

Beryland NG, Roze JN (1971) Use of correlation analysis in the regionalization of potential physical fields. Geomagn Aeronomy 11: 259 - 264

Beskow K, Granar L (1969) A computer program for 2-dimensional geophysical problems. Geoexploration 7: 137 - 152

Burg JP (1972) The relationship between maximum entropy spectra and maximum likelihood spectra. Geophysics 37: 375 - 376

Deeming TJ (1987) Band-limited minimun phase. In: Bernabini M, Carrion P, Jacovitti G, Rocca F, Treitel S, Worthington MH (eds): Deconvolution and inversion. Blackwell, London, pp 171 - 184

Goldman S (1955) Information theory. Prentice-Hall, New York, 385 pp

Holzmann (Golt'sman) FM (1971) Physical aspects of the statistical theory of the interpretation of observations. Izv Akad Nauk SSSR, Phys Solid Earth (7): 459 - 465

Holzmann (Golt'sman) FM (1976) The combination of observations in the identification of geophysical objects. Izv Akad Nauk SSSR, Phys Solid Earth 12 (7): 443 - 450

Holzmann (Golt'sman) FM (1982) Physical experiments and statistical inference. Leningrad, 191 pp (in Russian)

Holzmann (Golt'sman) FM, Kalinina TB (1983) Statistical interpretation of magnetic and gravitational anomalies. Leningrad, 248 pp (in Russian)

Holzmann (Golt'sman) FM, Protopopova AB (1968) Estimates of the maximum efficiency of geophysical interpretation. Izv Akad Nauk SSSR, Phys Solid Earth (3): 178 - 183

Holzmann (Golt'sman) FM, Kalinina TB, Besrukov SF (1988) Statistical evaluation of aggregate parameters of complex objects from a set of geophysical data. Izv Akad Nauk SSSR, Earth Physics, 24 (2): 139 - 147

Jackson DD, Matsu'ura M (1985) A Bayesian approach to nonlinear inversion. J Geophys Res 90 (B1): 581 - 591

Kalinina TB (1970) Quantitative estimation of the maximum efficiency of solutions of inverse problems in magnetic prospecting. Izv Akad Nauk SSSR, Phys Solid Earth (8): 487 - 494

Kalinina TB, Holzmann (Golt'sman) FM (1967) A statistical interpretation algorithm and its application in the solution of the inverse problen of magnetic exploration. Izv Akad Nauk SSSR, Phys Solid Earth, (7): 451 - 457

Kalinina TB, Holzmann (Golt'sman) FM (1971) Analytical comparison of methods of geophysical interpretation. Izv Akad Nauk SSSR, Phys Solid Earth (5): 320 - 325

Menke W (1989) Geophysical data analysis: Discrete inverse theory (revised ed). Academic Press, USA, 285 pp

Naudy H (1971) Automatic determination of depth on aeromagnetic profiles. Geophysics 36: 717 - 722

Olszak G, Peschel G (1971) Anwendung der Korrelationsanalyse bei der Komplexinterpretation. Z Angew Geol 17: 388 - 391

Peschel G (1971) Eine allgemeine Lösung für das Problem der quantitativen komplexen Interpretation geophysikalischer Messergebnisse. Z Angew Geol 17: 263 - 266

Peschel G (1972) Kriterien für die automatisierte komplexe Interpretation von geophysikalischen Profilen. Z Angew Geol 18: 500 - 502

Rao CR (1965) Linear statistical inference and its applications. John Wiley and Sons, New York, 522 pp

Shannon CE (1948) A mathematical theory of communication. Bell System Tech J 27: 379 - 423

Smith CR, Grandy WT Jr (eds) (1985) Maximum-entropy and Bayesian methods in inverse problems. D. Reidel, Dordrecht

Tarantola A (1987) Inverse problem theory. Methods for data fitting and model parameter esti-
    mation. Elsevier, Amsterdam, 613 pp
Tarantola A, Valette B (1982a) Inverse problems = quest for information. J Geophys 50: 159 - 170
Tarantola A, Valette B (1982b) Generalized non linear inverse problems solved using the least
    squares criterion. Rev Geophys Space Phys 20: 219 - 232

Chapter 6

ANALYTIC INVERSION

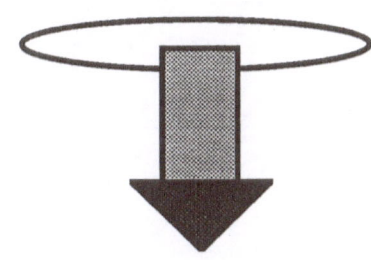

CHAPTER 6

ANALYTIC INVERSION

6.1 GENERAL PRINCIPLES

Analytical inversion would be the ideal approach to the inversion of any geophysical data (or any other physical data for that matter). Besides cost effectiveness (the computer time required for interpretation is often of the same order of magnitude as for the direct problem) analytical inversion can provide insight into the error propagation of the inversion. Even good approximations to analytical inversion may be useful as a part of a more general and complete scheme or computer program system for geophysical interpretation (Bott 1973).

Most geophysical fields, even of model bodies of simple shape, are too complicated for analytical inversion to be possible. But not even the most simple and evident possibilities have been used yet. In the following the use of some general principles, such as the combination of field components and algebraic manipulations, are demonstrated. The principles have also inherent properties, which may prove useful when new algorithms are developed for parallel computers. And the amazing capacities of new computer programs for algebraic manipulation certainly can be used to great benefit. New interesting and surprising improvements in the inversion procedures can certainly be found.

The combination of several components of a vector field at a single location is often a feasible simplification, which has useful interpretational features. The ideas of determining apparent resistivities in magnetotellurics and induction vectors in magnetometer array data analysis are good, almost classical examples of this principle. In the following the use of component combinations to transform a non-linear inverse problem into a simpler one through some characteristic examples. In these the whole or part of the inversion can be performed analytically.

6.1.1 Combination of Components

Let us first consider a simple example, the determination of the location and magnitude of a static magnetic dipole. The magnetic field of a dipole with moment $\mathbf{M}$ is:

$$\mathbf{H}_{dip} = \frac{3\,(\mathbf{M} \cdot \mathbf{R})\,\mathbf{R} - R^2\mathbf{M}}{R^5}, \tag{6.1}$$

where $\mathbf{R}$ is the vector from the location of the dipole $P_0(x_0, y_0, z_0)$ to the field point $P(x, y, z)$. If the location of P_0 is to be determined from field measurements, we have a strongly non-linear inverse problem, since the R^5 in the denominator is a function of the co-ordinates x_0, y_0 and z_0. The model parameter vector is $\mathbf{p} = (\mathbf{M}, \mathbf{R}_0) = (x_0, y_0, z_0, M_x, M_y, M_z)$. We would need data taken at three points or more to obtain a reliable inversion. The field expressions at each points would have a different R^5.

Algebraic manipulation leads to an equation of the 5th to 10th degree of the model parameters, which certainly cannot be solved analytically.

If we, however, have available measurements of three perpendicular components of the field at a single point, the expressions of the field components would have the same R^5 in the denominator. By dividing the components with each other, the strong non-linearity would disappear. The resulting inversion procedure (= the inversion operator) requires the solution of two equations of second degree and taking of one cubic root. The static dipole algorithm has been given by Hjelt (1977):

Step 1:   Compute   (6.2)

$$q_1 = M_y \cdot H_x - M_x \cdot H_y$$
$$q_2 = M_x \cdot H_z - M_z \cdot H_x$$
$$q_3 = M_y \cdot H_z - M_z \cdot H_y$$

$$g = q_3/q_2, \qquad\qquad h = -q_1/q_2$$

$$a = q_2(1 + g^2) - 3H_z \cdot (M_x + g \cdot M_y)$$
$$b = 2q_2 \cdot g \cdot h + 3H_x \cdot (M_x + g \cdot M_y) - 3H_z \cdot (M_z + h \cdot M_y)$$
$$c = q_2(1 + h^2) + 3H_x \cdot (M_z + h \cdot M_y)$$

Step 2:   Solve

$$a\,u^2 + b\,u + c = 0$$

Step 3:   Compute

$$v = g \cdot u + h$$
$$Q = M_x \cdot u + M_y \cdot v + M_z$$
$$S = u^2 + v^2 + 1$$

Step 4:   Get

$$z_0 = z - \left[\frac{3 \cdot u \cdot Q - M_x \cdot S}{S^{5/2} \cdot H_{x,\,obs}} \right]^{1/3}$$

$$x_0 = x - u \cdot (z - z_0)$$

$$y_0 = y - v \cdot (z - z_0)$$

If true field data are considered, the components of the magnetization vector are not known. So in fact three-component magnetic data taken at two field points are required to determine both the position **and** the strength of the dipole. It is easy to show that by determining the location from both data points assuming unit source strength, then a linear combination of the results give both **M** and the components of P_o. The accuracy of applying the algorithm to five data points located arbitrarily around a theoretical dipole gave a relative accuracy of $< 10^{-4}$ for the location and $< 10^{-3}$ for **M** (Fig. 6.1). The points should also be chosen so that **M** and **R** are not parallel or perpendicular to each other. Errors added to the data will naturally weaken the performance of the algorithm considerably (Fig. 6.2). The algorithm can be well used in EM inversion, since the direction of an induced secondary dipole can always be assumed to be known accurately enough. A further example of a combination of several field components is given in Section 6.2.1.

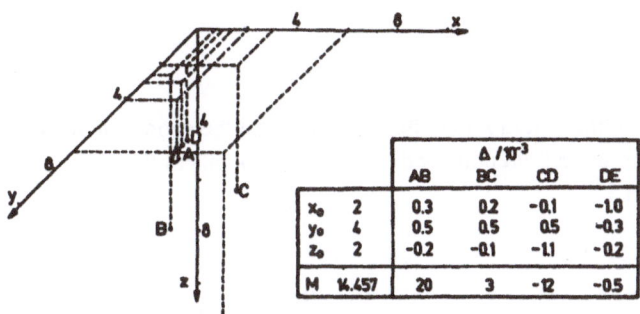

		$\Delta / 10^{-3}$			
		AB	BC	CD	DE
x_o	2	0.3	0.2	-0.1	-1.0
y_o	4	0.5	0.5	0.5	-0.3
z_o	2	-0.2	-0.1	-1.1	-0.2
M	14.457	20	3	-12	-0.5

Fig. 6.1. Determination of the location and magnetization of a static magnetic dipole, when the orientation of the dipole is known. Location of data points and table of errors (Hjelt 1975)

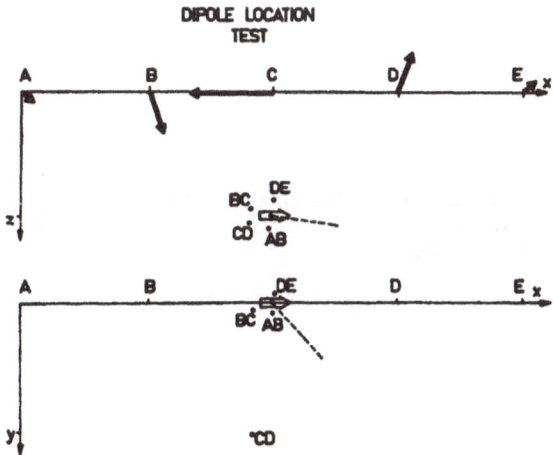

Fig. 6.2. Static magnetic dipole inversion with incorrect (assumed) dipole orientation produces large errors of the location (19 - 20 % of the total distance |R|). The algorithm fails whenever **R** and **M** are parallel or perpendicular to each other (Hjelt 1977)

6.2 EXAMPLES OF ANALYTIC INVERSION
(involving multicomponent techniques and complex functions)

6.2.1 Magnetometry

The magnetostatic field of a two-dimensional magnetized body can be written in a very compact form

$$F = I \cdot G \qquad (6.3)$$

by using complex entities, field F, amplitude factor I, and the geometric factor G, which is a function of the distances y between field point and characteristic points of the body (Hjelt 1976a, b). With the definitions in Fig. 6.3:

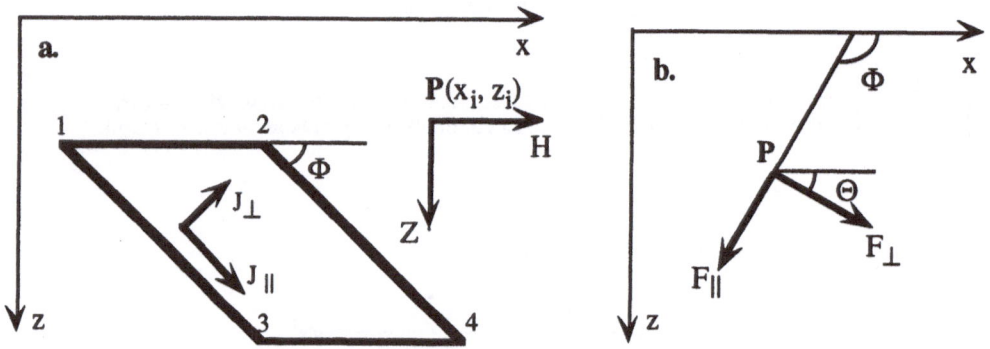

Fig. 6.3. The parameters of a thick magnetized plate and field components, when a complex number representation is used. The coordinates of the plate corners 1 - 4 are denoted (x_{10}, z_{10}) etc. (Hjelt 1976a, b)

$$F = H + j\,Z \qquad (6.4)$$

$$I = p + j\,q$$

$$y_i = (x_i - x_{io}) - j\,(z_i - z_{io})$$

For platelike bodies (Fig. 6.3)

$$G = \ln\left[\frac{(y_1\, y_3)}{(y_2\, y_4)}\right] \qquad (6.5)$$

and $I = -2 \sin \phi \cdot J$ with either $J = J_{||} + j J_\perp$ or $J = \alpha_\phi \cdot (J_x - j J_y)$.

α_ϕ is a dip-dependent complex number of unit length which is convenient to use for rotation of perpendicular components in the complex plane:

$$\alpha_\phi = e^{j\phi} = \cos \phi + j \sin \phi \tag{6.6}$$

For example for the borehole situation of Fig. 6.3 one would have:

$$F_b = F_{||} + j F_\perp = F \cdot \alpha_\Theta \quad \text{and} \quad F = F_b \cdot \alpha_{+\Theta} . \tag{6.7}$$

If the plate is thin compared to other dimensions, then

$$G_{thin} = \frac{1}{y_1} \cdot \frac{1}{y_4} \qquad \text{and} \quad I_{thin} = -2t \cdot J \tag{6.8}$$

with $t = d \sin \phi$ being the perpendicular thickness of the plate. Corresponding equations are valid for other, e.g. cylindrical bodies and also an extension of the expressions to gravimetry is possible.

Assume now, that two field points P_1 and P_2 are separated by a distance Δ_{21} (which is complex in the sense of definitions of y). For an infinitely deep thin plate:

$$F_1 = \frac{I}{y_1}$$

$$F_2 = \frac{I}{y_2} = \frac{I}{y_1 + \Delta_{21}} . \tag{6.9}$$

Dividing the two equations, one obtains immediately the inversion operator $L = \{y_1, I\}$:

$$y_1 = \frac{F_2 \Delta_{21}}{(F_1 - F_2)}$$

$$I = F_1 \cdot y_1 = F_2 \cdot y_2 . \tag{6.10}$$

This is an elegant and simple analytical equation for inverting magnetic field two-component data, the derivation of which was easy by using the complex notations.

The same principle can be applied further to obtain analytical inversion formulae for thin plates of finite depth extent and to a limited extent to thick plates. Measurements at three points are required for the finite thin plate case (Fig. 6.4) and the corresponding inversion formulae are:

$$y_1 = \frac{v + w_1}{2} \tag{6.11}$$

$$y_2 = \frac{v - w_1}{2}$$

$$I = -F_1 \cdot \frac{u}{w} ,$$

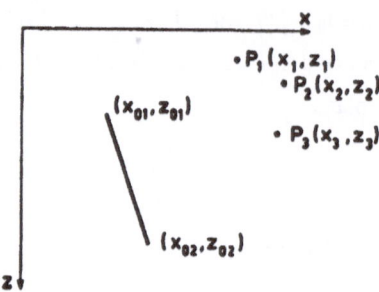

Fig. 6.4. Coordinates of a finite, thin magnetized plate. The three necessary measuring points can be located anywhere in the cross-section

where $w_1 = \sqrt{(v^2 - 4u)}$, $\quad w_2 = \sqrt{[v^2 - 4u(1 - F_1)]}$,

$$u = -\frac{c}{a}, \quad v = \frac{b}{a},$$

$$a = d_1 - d_2, \quad b = d_2 \cdot \Delta_{31} - d_1 \cdot \Delta_{21}, \quad c = F_2 \cdot F_3 \cdot \Delta_{21} \cdot \Delta_{31} \cdot \Delta_{32}, \quad d_1 = F_2 \cdot F_3 \cdot \Delta_{21},$$

$$d_2 = F_2 \cdot F_3 \cdot \Delta_{31}.$$

and $\quad y_{ij} = x_i - x_{oj} - j \cdot (z_i - z_{oj})$

with $i = 1, 2$ and $j = 1, 2, 3$.

For a thick plate F_i is replaced by $F_i = 1 - E_i = 1 - e^{(F_i/I)}$. The problem with this case is that the magnetization I has to be known or assumed, before the ln-function in the thick plate expression can be inverted and the plate parameters can be determined. This would naturally call for an iterative approach, where the geometrical parameters are determined using the analytical inversion expressions and the magnetization is determined by a linear procedure (since I appears in the original forward field expression linearly!).

One of the definite advantages of analytical inverse operators is, that they can be used to study and demonstrate the sensitivity of the inversion both to data errors and anomalous fields caused by other than the causative bodies. Hjelt (1976b) has given such an example for the vertical magnetized thin plate model. By applying the operator to four data point pairs, which include a random error component, the scatter of the cluster of resulting plate edge locations (crosses in Fig. 6.5) is indicative of the error sensitivity of the operator. Random errors in the field itself affect the inversion much more than errors in measuring point locations. When the field data contain both random and a systematic zero level error, the inverted locations are more scattered. They seem, however, to group somewhat more systematically around a parabolic trace, which seems to be typical for zero level error.

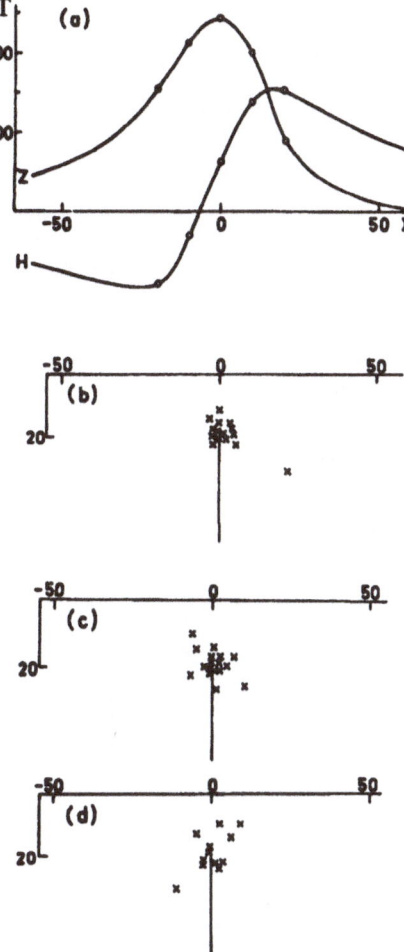

Fig. 6.5. a. The vertical (Z) and horizontal (H) components of anomalies. **Circles** are the data points used in inversion. **Crosses** in (b) - (d) correspond to the average of position determinations from four of the point pairs. (Hjelt 1976 b)

Hjelt (1976a) has further suggested that the thin, infinitely deep plate inversion can be applied to arbitrary magnetic anomaly profiles to obtain qualitative information concerning the number and type of magnetized bodies in the cross-section. If the infinitely deep thin plate inverse operator is applied, then the typical patterns of the location of the upper edge can be listed approximately as in Table 6.1.

Example of the loops occurring for multiplate anomalies are shown in Fig. 6.6 Since the inverse operator, as any inverse formula, is very sensitive to data errors, the possibility of applying this principle in practice is limited (cf. borehole example). When the operator is used iteratively to a multiplate profile, the inversion proceeds rather favourably (Fig. 6.7). Such an approach would definitely have virtues as a part of an inversion system operated on a parallel computer.

TABLE 6.1 Typical upper edge location patterns

"Data" field type	Pattern of inversion
Plate, ∞ depth, + Zero level	Parabola with its apex at the edge of the plate
Plate, ∞ depth, + gradient	Closed loop encircling the edge of the plate
N plates, ∞ depth,	N-1 closed loops encircling the edge of the plate
Plate, finite depth	Bell-shaped curve above the edge of the plate
Thick plate, ∞ depth	Circular loop indside the upper part the plate

a.

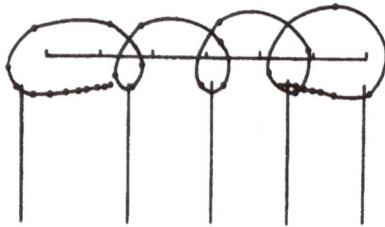

b.

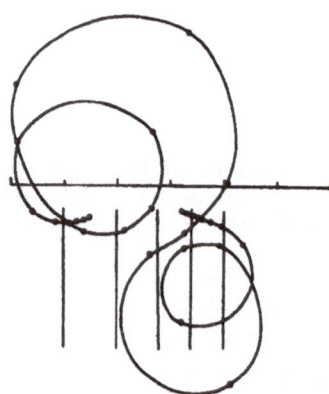

Fig. 6.6. Location of plate edges obtained by analytical inversion with the thin, infinitely deep plate operator. The "data" profile is caused by five parallel, vertical plates. **a.** Widely separated plates; **b.** densely located plates. (Hjelt 1976 a)

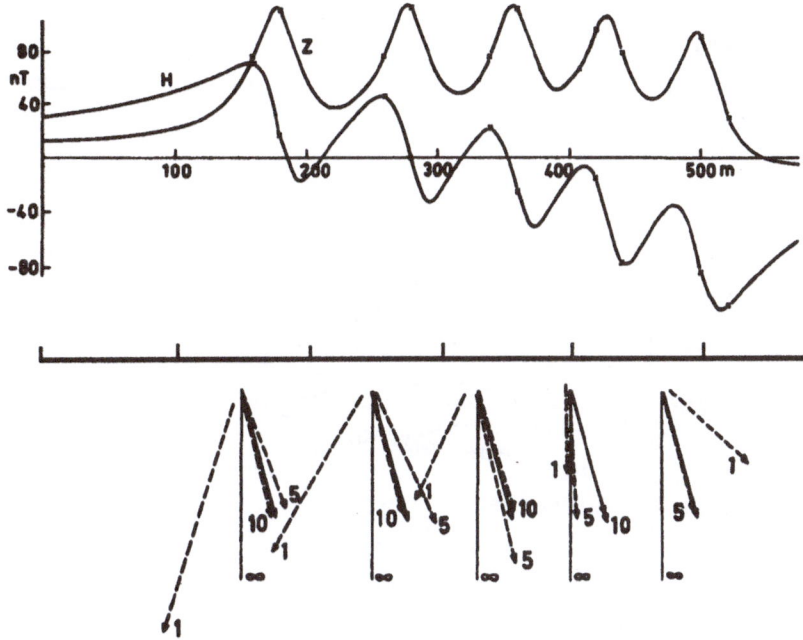

Fig. 6.7. Inversion of a multiplate, two-component magnetic anomaly profile (**above**) using the analytic thin plate inversion operator iteratively to the data points indicated by **crosses**. The **dashed arrows** represent the inverted magnetization vector and **numbers** refer to iteration cycle (**below**). The true model is given by the **full line arrows**. (Hjelt 1976 a)

A multibody (long) anomaly profile can be inverted more successfully by applying the plate algorithm iteratively. A few points (the number depends on whether an infinite or a finite plate model is used) around each anomaly maximum is used to produce an estimate for the corresponding plate. Every time after having determined a new plate, its effect is subtracted from the data at the other bodies. The Fig. 6.7 (from Hjelt 1976a) shows how the magnetization vectors quickly approach their correct location, direction and amplitude of magnetization, when the five plates are well apart from each other (about three to five times the depth of the upper edges in this case). After ten iterations the magnetization and location of the plates cannot be distinguished from the true values within drawing accuracy. When the plates are closer to each other (Hjelt 1976a gives an example ons the magnetization and location of the plates cannot be distinguished from the true values with five plates separated by only 1.3 - 2 times depth), the anomalies become less distinct and the iteration converges, but more slowly. By constraining the magnetization vectors to keep their direction parallel to the direction of the magnetizing geomagnetic field, good convergence is obtained even for plates close to each other.

The analytical plate inversion has been tested also on multicomponent magnetometer field data. The inversion result, shown in Fig. 6.8, was obtained by iterative use of the finite plate algorithm and proved successful when considered in vertical cross-section (bottom part of the Fig. 6.8 b). The lower of the plates simulates well the ore body, known at that time to the depth of slightly more than 400 m, but having obvious extensions according to the boreholes. The model plate ends a few tens of metres below the borehole R84 in accordance with the pattern of borehole magnetometer vectors (not shown). The upper model plate is weakly magnetized and simulates some local weaker borehole intersections (dashed areas).

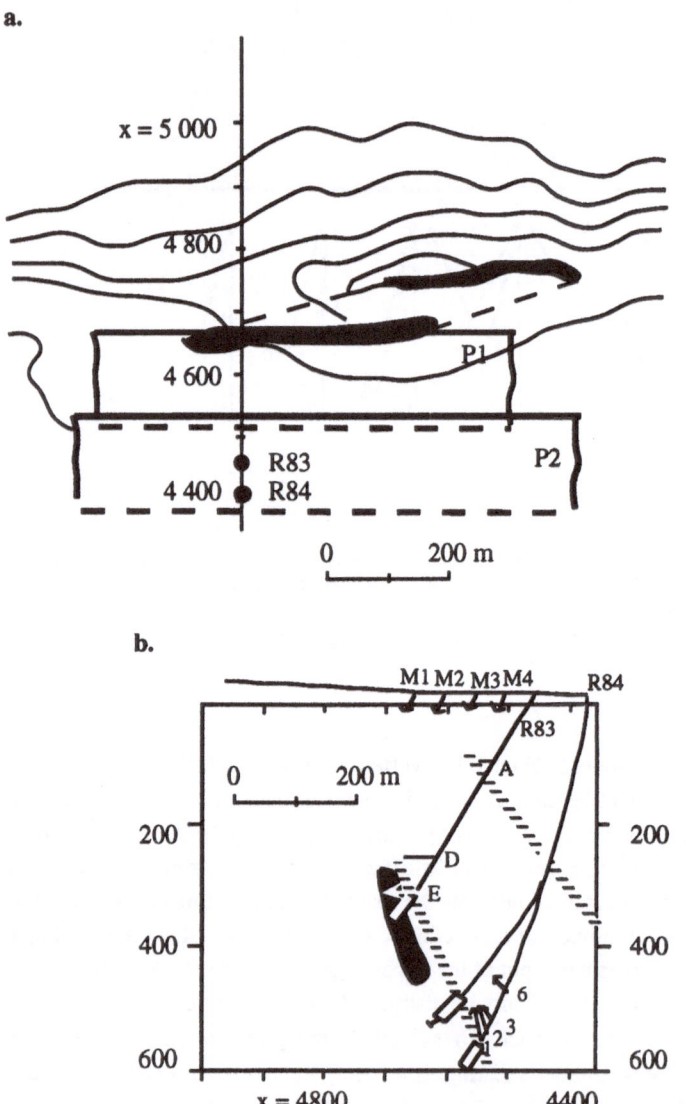

Fig. 6.8. Application of the analytical plate inversion algorithm to real data. **a.** Surface map with contours of total anomalous magnetic field and projections of the 2D plate models. **Black areas** cross-sections of the ore body at 0 and 400 m depths. **b.** Vertical cross-section with boreholes, magnetic data vectors, the ore body (**black**), areas with strongly magnetized borehole samples (**white areas** around boreholes) and the plate models (hatched). (Hjelt 1975)

The algorithms were tested extensively to both borehole and surface data, but for reliable results the data used for inversion had to be selected and/or smoothed. This is again a reminder of the error sensitivity especially of analytical inversion procedures. Figure 6.8 b. also shows that the strongest magnetized body, the ore, is far from being two-dimensional. The profile, for which the vertical section is shown, passes over the lower parts of the ore body, which seen from the surface resembles more a two-dimensional situation. This explains the reasonable success of the plate model for the deep parts of the ore.

6.2.2 EM Magnetic Dipole and an Infinitely Well Conducting Half-Plane

The forward problem of calculating inductive electromagnetic fields of infinitely well conducting bodies can sometimes be solved analytically. Even if the model of an infinitely well conducting body cannot be used to determine the conductivities of geological structures, they can be used for geometrical determinations to locate the bodies. Their analytical forms may be very useful as part of a more complete inversion algorithm of EM inductive methods.

For example the case for a half-plane located in the field of a harmonically varying magnetic dipole is considered. The forward problem has an analytical solution (Sommerfeld 1897; Grant and West 1965). The complete field expressions for arbitrarily oriented transmitter and receiver dipoles are, despite the analytical form, still rather involved. They are considerably simplified, if double-dipole systems are considered (Hjelt 1968). Hjelt (1977) has shown, that for a proper combination of dipole-dipole geometries, the inverse problem of finding the location and dip of the half-plane can be solved (almost) analytically. Double-dipole geometries are useful system approximations for most of the airborne EM dipolar configurations.

Let us consider the half-plane model (Fig. 6.9). It can be described by a three-component parameter vector $\mathbf{p} = (x_0, h$ and $\Phi)$, where x_0 = horizontal location of the edge of the plane, h = vertical depth to the edge of the plane and Φ = dip of the plane with respect to the horizontal location. The forward problem to the field of an infinitely well conducting half-plane in a dipolar source field was originally solved by Sommerfeld (1897). It was reformulated for use in geophysics by West (Grant and West 1965). The induced secondary field is obtained by double derivation of Green's function containing nothing more than inverse tangent function and square roots (Fig.6.10). The location of the transmitter (source dipole) is denoted by P_0 and the receiver (field point) by P. P_2 is the mirror point, where the "image" dipole simulating the inductive currents induced in the half-plane is placed. It is elementary, albeit tedious, to extend the field expression once the field geometry is specified.

Using the notations of Fig. 6.10 we have:

$$\mathbf{H}_p = \frac{1}{4\pi} \nabla_p (\mathbf{m} \cdot \nabla_o G), \qquad \text{with } G = T_1 - T_2$$

$$T_i = \frac{1}{\pi R_i} \text{atan} (R_i / g_i), \, i = 1, 2$$

$$g_1 = \sqrt{2(\rho \rho_o + x \, x_o + y \, y_o)}$$

$$g_2 = \sqrt{g_1^2 - 4x \, x_o} \; .$$

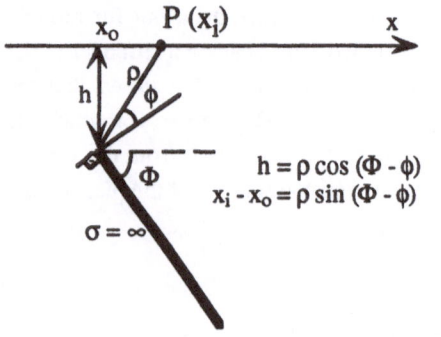

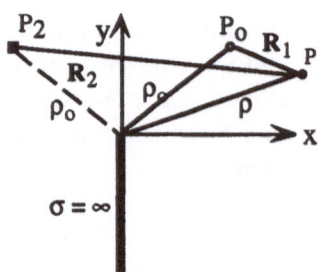

Fig. 6.9. Definition of the half-plane model parameters

Fig. 6.10. Cartesian and polar coordinates used in calculating the secondary EM field of an ∞ well conducting half-plane in the inducing field of a magnetic dipole

For double-dipole geometries P approaches P_0 and the primary field approaches mathematical infinity. The secondary field, however, remains finite and by careful manipulation of the equations the singularities of the field can be avoided. Using a notation (Hjelt 1968), where a superscript denotes the direction of the source dipole and the subscript the direction of the receiver dipole (axis of the receiving loop), the configurations of Fig. 6.11 correspond from left to right to vertical coaxial, crossed coil ("whale-tail") and horizontal coplanar EM systems.

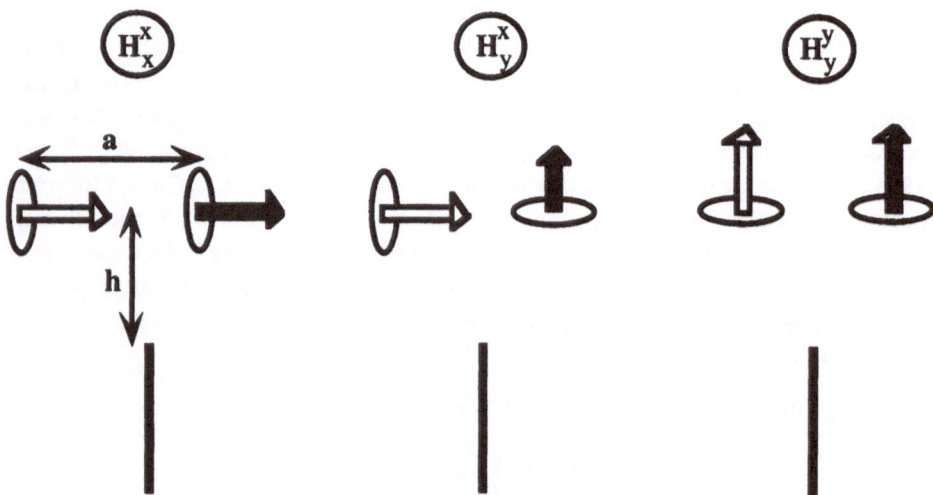

Fig. 6.11. Notations used for three double-dipole systems. **Open arrow** source; **solid arrow** receiver dipole

The fields of these three systems can be written (Hjelt 1968) as:

$$H_x^x = A \cdot (K_1 \cdot \sin^2 \Phi - K_2 \cdot \sin 2\Phi + K_3 \cdot \cos^2 \Phi)$$

$$H_y^y = A \cdot (K_1 \cdot \cos^2 \Phi + K_2 \cdot \sin 2\Phi + K_3 \cdot \sin^2 \Phi) \qquad\qquad (6.12)$$

$$H_y^x = H_x^y = A \cdot [(K_1 - K_3) \cdot \sin 2\Phi - 2 \cdot K_2 \cdot \cos 2\Phi]/2 \,,$$

where

$$K_1 = \frac{5}{6} - 2F - \sin \varphi \cdot (\sin \varphi - 0.5)$$

$$K_2 = \cos \varphi \cdot (\sin \varphi - 0.5)$$

$$K_3 = \sin \varphi \cdot (\sin \varphi - 0.5) - \frac{1}{6} - F$$

$$F = \frac{\pi - 2\varphi - \sin 2\varphi}{4 \cos^3 \varphi} \qquad \text{and} \qquad A = \frac{M}{16 \, \pi^2 \rho^3} \,.$$

In the expressions the cylindrical co-ordinates ρ and φ (cf. Fig. 6.10) have been used. The model parameter vector thus can be written as $p = (\rho_0, \varphi$ and $\Phi)$ instead of the original form $p = (x_0$, h and $\Phi)$. These simple field expressions remain valid also for finite transmitter-receiver separations, a, as long as the quotient a/h remains < 0.3.

The field expressions contain three types of functions:

1. The amplitude factor A, which depends only on the distance ρ.

2. The functions K_1, K_2 and K_3, each one depending only on the direction parameter φ.

3. Each factor K is multiplied by a trigonometric function depending on the dip Φ only.

All functions and field components depend non-linearly on the model parameters ρ_0, φ and Φ (or x_0, h and Φ). In order to develop inverse expressions in analytical form, the three field components have to be combined so, that the new expressions are functions of a single parameter only. For example the combinations

$$(H_x^x)^2 + (H_y^y)^2 + 2 \, (H_y^x)^2 \quad \text{and} \quad (H_x^x + H_y^y)^2 \,,$$

are such that the dip parameter Φ is eliminated. Dividing these expressions by each other eliminates the amplitude A and the result, T, depends only on the direction parameter φ. Determining T from the measured data, a non-linear equation to determine φ is obtained. Although no single analytical inversion of this equation has been found, it does not depend on the dip of the plate. A suitable functional (e.g. polynomial) approximation can be developed for this inversion. Using the direction parameter solution, two further amplitude-independent functions, P and R, are calculated from data and the K-functions, respectively. The distance is obtained by dividing any field component, computed for unit amplitude, with corresponding measured field component. The resulting algorithm of inverting EM double-dipole data is:

Step 1: Determine $\hspace{11cm}$ (6.13)

$$T = \frac{(H_x^x)^2 + (H_y^y)^2 + 2\,(H_y^x)^2}{(H_x^x + H_y^y)^2} \qquad \text{from the data}$$

Step 2: Solve φ_s from:

$$K_1^2 + 2\,K_2^2 + K_3^2 - T\cdot(K_1 + K_3)^2 = 0$$

Step 3: Compute from the data and using the solved φ_s:

$$P = \frac{2\cdot H_y^x}{H_x^x - H_y^y} \qquad\qquad R = \frac{K_1 - K_3}{2\,K_2}\bigg|_{\rho_s}$$

Step 4: Solve

$$\Phi_s = \frac{1}{2}\,\text{atan}\left[\frac{1 - P\cdot R}{P + R}\right]$$

Step 5: Solve

$$\rho_s = \left[\frac{M}{16\,\pi^2}\,\frac{2\cdot H_y^x(A = 1,\ \Phi_s,\ \varphi_s)}{H_y^x\ (\text{measured})}\right]^{1/3}$$

Step 6: Transform the cylindrical model parameters into rectangular ones:

$$x_0 = x_i - \rho_s\cdot\sin B$$

$$B = \Phi_s - \varphi_s$$

$$h = \rho_s\cdot\cos B$$

In a test with theoretical data the algorithm was applied sequentially to all points along a profile of length varying from 2.5 to 6 units. With a point separation of 0.25 units this corresponded to 10 to 25 data points. The test model was $p = (x_0 = 0, h = 1, \Phi = 60^o)$. When Gaussian errors, 0.01 % of the greatest amplitude were added to the "data", the location was determined with an accuracy better than 0.1 units and the greatest error of dip was only a few degrees. When the data error was increased to 1 %, depth determination was mostly affected, whereas the accuracy of horizontal location and dip determination remained almost the same (Hjelt 1977).

REFERENCES

Bott MHP (1973) Inverse methods in the interpretation of magnetic and gravity anomalies. In: Bolt BA (ed): Methods in computational physics vol. 13, Academic Press, USA, pp 123 - 162

Grant FS, West GF (1965) Interpretation theory of applied geophysics. McGraw-Hill, USA, 584 pp

Hjelt S-E (1968) On the half-plane model in double dipole electromagnetic prospecting. Acta Polytechn Scandin, Physics incl Nucleonics Series, Ph 60, 41 pp

Hjelt S-E (1975) On the interpretational advantages of many-component measurements. Department of Geophysics, Univ of Oulu, Contribution 48, Oulu, 15 pp

Hjelt S-E (1976a) A new approach to magnetic profile interpretation. Geophys Prosp, 24: 1 - 18

Hjelt S-E (1976b) A new compact formulation of the two-dimensional magnetic interpretation problem. Geoexploration14: 1 - 20

Hjelt S-E (1977) Many-component measurement as a means of simplifying EM interpretation problems. Acta Geod Geoph Montan Acad Sci Hung 12: 301 - 309

Sommerfeld A (1897) Über verzweigte Potentiale im Raum. Lond Math Soc Proc 28: 395 - 429

ADVANCED INVERSION METHODS

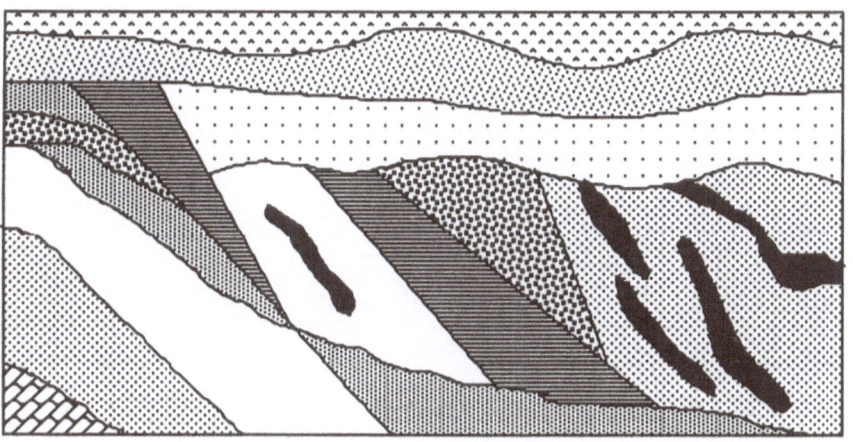

The real EARTH is beautiful - it is complex !

CHAPTER 7

ADVANCED INVERSION METHODS

All the inversion techniques described so far have more or less required some degree of selection of a model to describe the geometrical features of the subsurface. There exists, however, a wealth of more advanced approaches both in the geophysical and the mathematical literature. In these either the effect of the subsurface is simulated by a set of primary source distributions of the corresponding geophysical field or the inversion is aimed directly at the basic differential equation in question. It is typical for these methods, that they have a mathematically very advanced background. Although they are of general mathematical form, the methods sometimes are closely tied to a certain type of geophysical field and their development is not always on a practical level to transform them directly for an application to other methods. The versatility of eg. the SVD technique or non-linear optimization is far better developed at the moment. In the following some examples have been selected to highlight the possibilities of these methods.

7.1 FUNCTIONAL ANALYTIC METHODS

Functional analysis is an important mathematical tool when dealing with various types of potential fields. Everybody is, for example, familiar with the importance of the theory of complex functions for 2D gravity and magnetic field problems. As a loose definition one could say that most of the inversion methods based on functional analysis have some connection with the continuation of fields. Also multipole expansions of field can be considered as belonging to this category. These methods require typically iterative solutions, where starting with the field described by the data one tries to extrapolate the field closer and closer to the originators of the field. The originators can be described by firstly simple geometrical bodies (as has been the case in all previous chapters), secondly by simple excitors like points, lines, dipoles or multipoles or finally by functional approximations of boundaries between structures of the subsurface, which differ in their physical properties from the surroundings. All these techniques are numerically instable, because the ill-posed original problem (= a small change in the boundary condition, e.g. the field measured on the Earth's surface) leads to large or infinite changes in the field close to its source. The monographs by Lavrentiev (1967), Tikhonov and Arsenin (1977), among others, describe the question of ill-posedness thoroughly, albeit in a mathematically demanding way.

7.1.1 Methods of Singular Points

When the originators (sources) of the field are described by simple excitors like points, lines, dipoles or multipoles, one often speaks of the method of singular points. This has its natural mathematical origin in the fact, that most geophysical fields depend on $1/R^n$, where $R = r - r_0$ with r being the position vector of any field (data) point and r_0 the location vector of the source point. When the field is extrapolated during the inversion process towards the sources r approaches r_0, consequently R approaches 0 and the field becomes singular (approaches ∞). In the method of singular points, this feature is deliberately put to use. For singular lines (branch lines, say) the problem of multi-valued functions also enters the scene. For geometrical bodies or functional descriptions of structural boundaries the singularity is often less serious, since in order to calculate the field one has to integrate over the surface of the body. Some extremely simple bodies may contain singular points. A commonly encountered example is a half-plane, where the fields at the tip of the plane behave singularly. The approach of a body surface can, however, in most cases be identified from the increasing instability or oscillation of the continued field (cf. example in Sect 7.2).

7.1.2 Method of Tightening Contours

In the method of tightening contours the form of a body with excess material properties is constructed iteratively. The surface of a (in this example, the 2D body) is described by $\rho = f(\phi)$ (cf. Fig. 7.1). The functional to be minimized is for example:

$$S(f) = \|E_y(f) - E_{h,y}\|^2 ,\qquad\qquad(7.1)$$

where E_y is the horizontal component of the geoelectric field. Regularization can be added to the

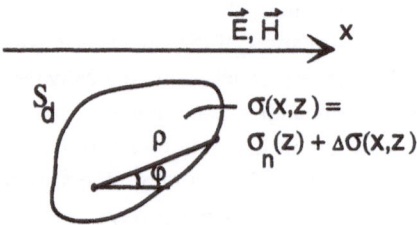

Fig. 7.1. Definition of the model. The boundary S of the body with excess conductivity $\Delta\sigma(x, z)$, is defined by polar co-ordinates ρ and ϕ in relation to a specified origin

solution, whereby a stabilizing functional T(f) is added to S(f). The functional T is preferably chosen so that the contour is as smooth as possible or close to a given function f_o, i.e.:

$$T(f) = \|f - f_o\|^2 = \int_{-\pi}^{\pi} (f - f_o)^2 \, d\varphi \,.$$ (7.2)

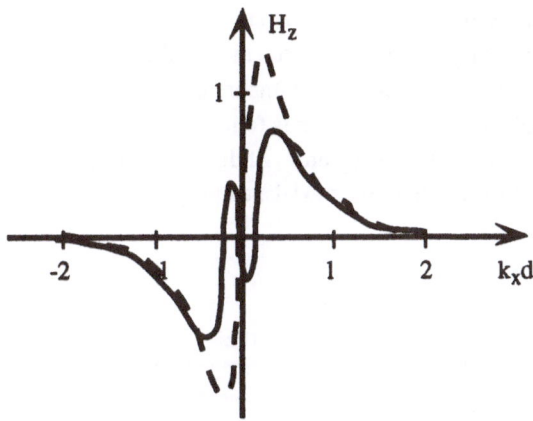

Fig. 7.2. The vertical magnetic field of a horizontal circular conducting cylinder. Conductivity is 10 000 times the conductivity of the surroundings. The ratio of radius to depth, **a/d** is 0.25 and the **wavelength** of the EM plane wave is $\lambda = 25$ d. The fields are normalized with the horizontal field of the homogeneous Earth. (Berdichevsky and Zhdanov 1984)

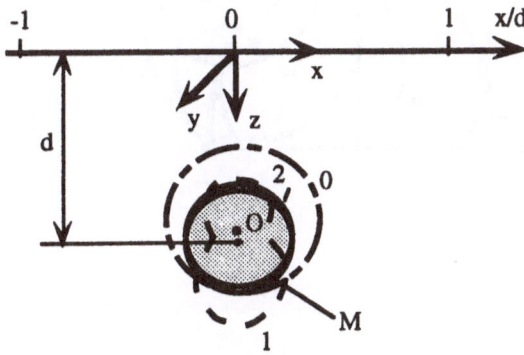

Fig. 7.3. The cylinder model inverted by the method of tightening contours. Models: **0**: initial (with centre at 0); **1**: final without regularization; **2**: final with regularization; **M**: original model. (Berdichevsky and Zhdanov 1984)

The regularized objective function to be minimized is:

$$S_\alpha(f) = S(f) + \alpha \cdot T(f). \tag{7.3}$$

The practical implementation of the technique has shown that the origin of the subsequent models has to be controlled in some way, so that it cannot "run wild", that is migrate to a location outside the convergence area of the iteration. The technique is demonstrated by two examples given by Berdichevsky and Zhdanov (1984). In the first case a cylindrical horizontal conductor is re-constructed starting from a profile of the vertical component of the geomagnetic field (Fig. 7.2). Although the initial model is chosen close to the true model (the initial cylinder encompasses the true model), the example in Fig. 7.3 shows clearly the important role played by regularization.

In the second example the flow of iteration is followed for a conducting half-cylindrical model. Fig. 7.4, taken again from Berdichevksy and Zhdanov (1984), concentrates on showing the importance of the control of the origin of the models, although no escaping situation is shown.

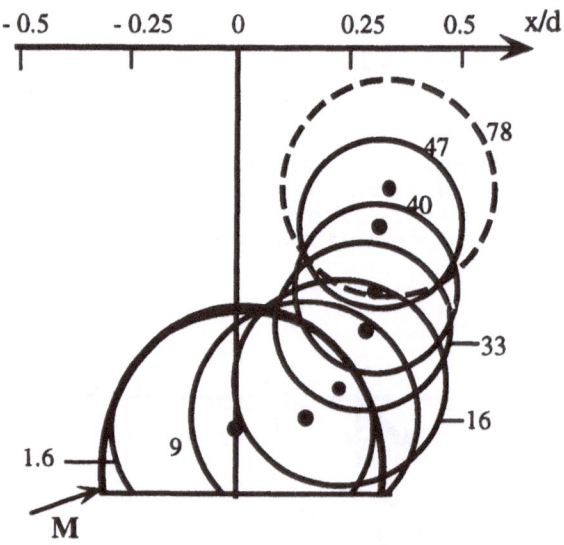

Fig. 7.4. Determination of the form of a conducting half-cylinder by the method of tightening contours. The conductivity of the inclusion is 100 times the surroundings. **Dashed line** initial model; **full lines** successive iterations; **black dots** the correct centers of the models; **numbers** on the contours: % errors; **M** original model (after Berdichevsky and Zhdanov 1984)

7.1.3 Method of Finite Functions

In the method of finite functions, the boundary of a region having a physical parameter differing from the surrounding subsurface is not prespecified as in most examples treated earlier. Instead the parametrization is made by selection of suitable functions to describe more complicated boundaries of geological structures of the subsurface. Zhdanov and Golubev (1983) has given an example from electrical modelling. The excess conductivity of the region is given by:

$$\Delta\sigma\,(x, z) = A \cdot f_1(x) \cdot f_2(z) . \tag{7.4}$$

Multiparameter functions f are defined by:

$$f(x) = \begin{cases} 0 & x \notin [a,b] \\ A \cdot \Phi_{\alpha\beta}(t) & x \in [a,b] \end{cases} \tag{7.5}$$

with $\quad \Phi_{\alpha\beta}(t) = (1 - t)^\alpha \cdot (1 + t)^\beta$

and $\quad t = -\dfrac{i}{\pi}\ln\left\{\dfrac{(1 + k^*)(e^{i\pi\tau} - k)}{(1 + k)(1 - e^{-i\pi\tau})}\right\} \qquad\qquad k = p + i\,q; \quad |k| < 1$

$$\tau = -\dfrac{2x - (a + b)}{b - a} \qquad\qquad -1 \le \tau \le 1$$

When x varies within the original boundaries from a to b the parameter τ performs a linear mapping and varies along the line [-1, 1]. The parameter t furthermore performs a mapping from [-1, 1] to [-1, 1]. The total number of parameters is five (A, α, β, p and q; t and τ have the role of intermediate parameters only). The function f naturally depends also on a and b, the limits of the area (line) of variation of the profile co-ordinate x. The descriptive properties of the function is demonstrated by a few selected examples in Fig. 7.5.

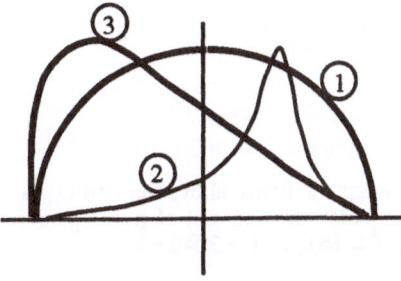

Fig. 7.5. The versatility of the described finite function. **1** $\alpha = 0.5$; $\beta = 0.5$; $p = 0$; $q = 0$; **2** $\alpha = 2.0$; $\beta = 2.0$; $p = 0.5$; $q = 0.5$; **3** $\alpha = 2.0$; $\beta = 0.5$; $p = 0$; $q = 0$. (Zhdanov and Golubev 1983)

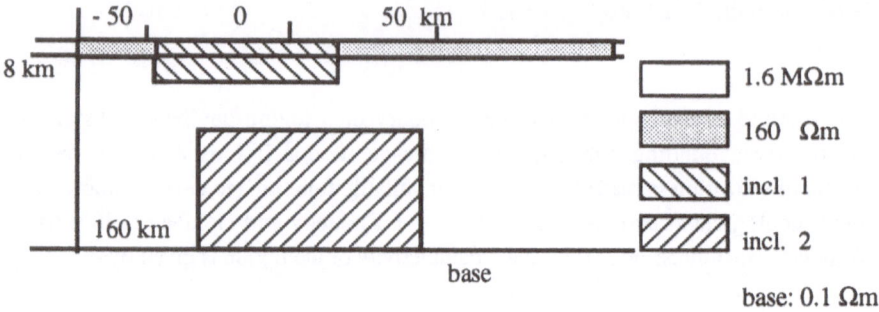

Fig. 7.6. A three-layer subsurface model with two inhomogenous conductivity inclusions. (Berdichevsky and Zhdanov 1984)

a.

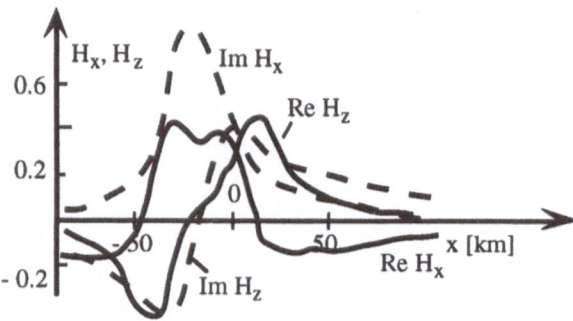

b.

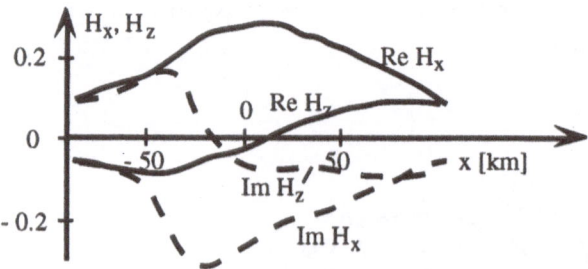

Fig. 7.7. The vertical and horizontal magnetic fields above the Earth structure model in Fig. 7.6. **Full line** in-phase parts; **dashed** out-of-phase parts (after Berdichevsky and Zhdanov 1984) **a.** T = 16 s; **b.** T = 2048 s

Zhdanov has used the method in studying the inversion of the vertical and horizontal components of the geomagnetic field variation at two periods, $T_1 = 16$ s and $T_2 = 2048$ s. The inducing geomagnetic field was assumed to be a plane E-polarized wave. The model of the subsurface is a three-layer Earth which contained two inclusions, the excess conductivity of which was assumed to vary horizontally (Fig. 7.7). The size of the upper inclusion is 8 x 61 km and that of the lower one 100 x 76 km. The variations were modelled by one-dimensional finite functions with $\alpha_{1,2} = 1$ and $\beta_{1,2} = 1$. The data [H_x (x; T_1 and T_2) and H_z (x; T_1 and T_2)] are shown in Fig. 7.7.

a.

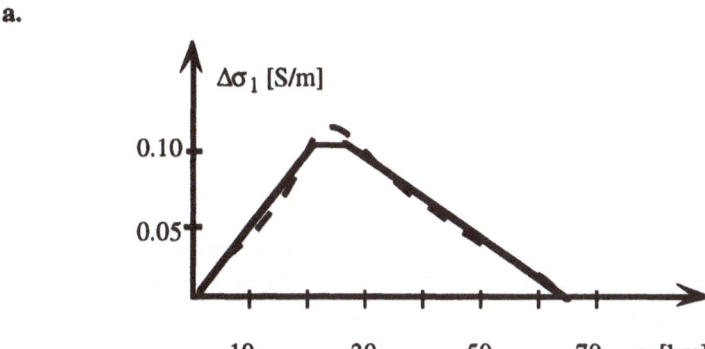

b.

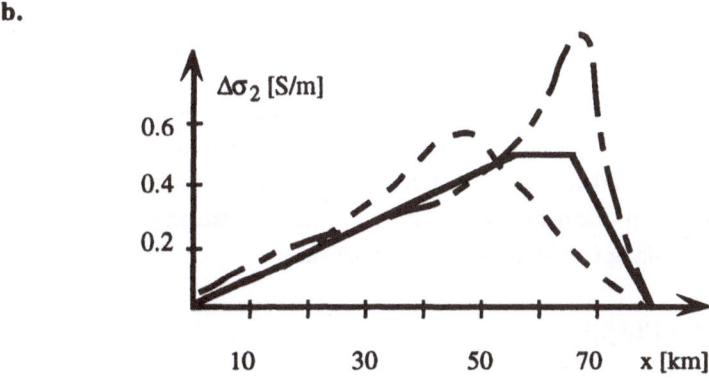

Fig. 7.8. Excess conductivity of the inclusions. Model with $\alpha = 1$ and $\beta = 1$. **Full line:** original model; **dashed lines:** inversion model. **a.** Surface (**upper**) inclusion. **b.** Lower inclusion (Berdichevsky and Zhdanov 1984)

Using non-linear optimization, the electrical conductivity of the upper inclusion $\Delta\sigma_1$ was reasonably well modelled, whereas the optimization of conductivity of the lower inclusions $\Delta\sigma_2$ seems to be unstable (a tendency to oscillate around the true distribution; Fig. 7.8).

7.2 CONTINUATION OF FIELDS

The mathematical continuation of a function from a region 1 to another region 2 is possible, whenever (a) the function is analytic, (b) the mathematical union of the regions is non-empty and (c) the singular points of the function can be isolated. The continuation of (static) potential fields can be considered as a classical technique within geophysics. The basic principle is well covered in the first volume on geophysical interpretation by Grant and West (1965), in some later specialized books (see Zhdanov 1988; Zidarov 1990). For potential methods, e.g. the gravity field Δg, at a height z above the boundary S can be written:

$$\Delta g_z(\mathbf{r}) = \frac{|z|}{2\pi} \int_{-\infty}^{\infty} \frac{\Delta g_0(\mathbf{r}_0)}{|\mathbf{r} - \mathbf{r}_0|^3} dS \,. \tag{7.6}$$

In this expression Δg could be replaced by any geophysical field satisfying the Laplace equation.

If the gravity field Δg_0 is known on the surface S, the field can be continued away from the source simply by calculating the integral in the equation. This is called *upward continuation*. Integration is a smoothing mathematical operation, whence upward continuation smooths the data or in other terms filters out the short wavelength features of the field. An important practical application is to tie together magnetic measurements taken from airplanes or satellites at different altitudes into a common (higher lying) level. Even ground and airborne data can be successfully levelled by this technique.

If Δg_z represents the known (measured) gravity field, and one wishes to determine the unknown Δg_0, it becomes necessary to solve the integral Eq. (7.6). This is an inherently unstable (ill-posed, see Lavrentiev 1967; Tikhonov and Arsenin 1977) problem. Since Δg_0 is closer to the sources than Δg_z, this solution, if calculable, will be the *downward continuation* of the gravity field. It can formally be solved by taking Fourier transforms of both sides,

$$F\{\Delta g_z\} = e^{-\sqrt{p^2 + q^2}\, z} \cdot F\{\Delta g_0\} \tag{7.7}$$

solving for the transform of Δg_0:

$$F\{\Delta g_0\} = e^{+\sqrt{p^2 + q^2}\, z} \cdot F\{\Delta g_z\} \,, \tag{7.8}$$

and by taking the inverse transform to obtain Δg_0 itself

$$\Delta g_0 = F^{-1}\left\{ e^{+\sqrt{p^2 + q^2}\, z} \cdot F\{\Delta g_z\} \right\}. \tag{7.9}$$

This latter transform does not exist, unless the spectrum of the transform of the data Δg_z attenuates more rapidly as a function of p and q than the exponential in Eq. (7.9). This is not true in the general case, whence the expression:

$$e^{+\sqrt{p^2+q^2}\,z}.F\{\Delta g_z\}, \tag{7.10}$$

has to be smoothed artificially. The various regularization techniques (see e.g. Tikhonov and Arsenin 1977) for doing this are expedient for successful downward continuation. No regularization procedure will allow the continuation to be taken across any surface, containing the sources of the field. The formal solution by integral transformation demonstrates clearly that downward continuation, if successful, performs low-pass filtering of the data, attenuating the long wavelengths and amplifying the shorter ones (inclusive data errors). Of the two continuation processes only downward continuation is a genuine inverse procedure.

To demonstrate the behaviour of downward continuation, two examples from electromagnetic field theory (Berdichevsky and Zhdanov 1984) are shown. Firstly the in-phase and out-of-phase components of the vertical magnetic field above an infinitely well conducting strip in a conducting half-space (Fig. 7.9) and a plane wave field are shown at various levels above the strip (Fig. 7.10). The last two profiles are located in the source region (at the tip of the strip and crossing the strip). Regularization and the finite conductivity of the surrounding ($\lambda_n = 100$ d) smooth the oscillatory behaviour of the field at these depths.

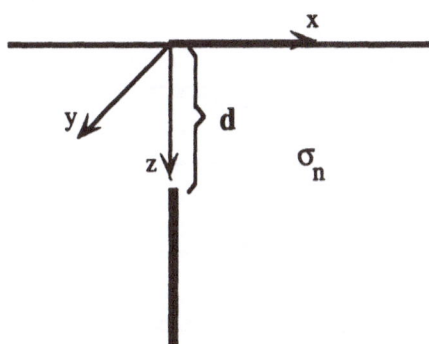

Fig. 7.9. The perfectly conducting, vertical strip in a conducting half-space. The strip is located in the plane x = 0. (After Berdichevsky and Zhdanov 1984)

The second example shows how the smoothed in-phase (real) and out-of-phase (imaginary) parts of the magnetic field components are focussed at the vertices and the upper surface of the rectangular (square) conductive inclusion, whereas the horizontal electrical field bends smoothly around the inclusion. It is evident that in this situation the magnetic field contains more information about the inclusion than the electric field. The model is defined in Fig. 7.11 and the inverted contour lines in Fig. 7.12.

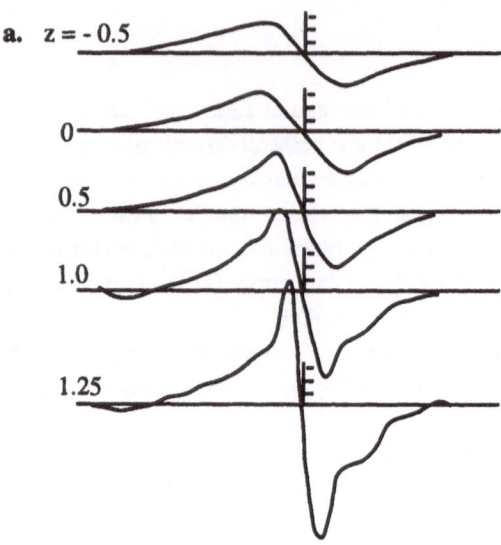

a. z = - 0.5

0

0.5

1.0

1.25

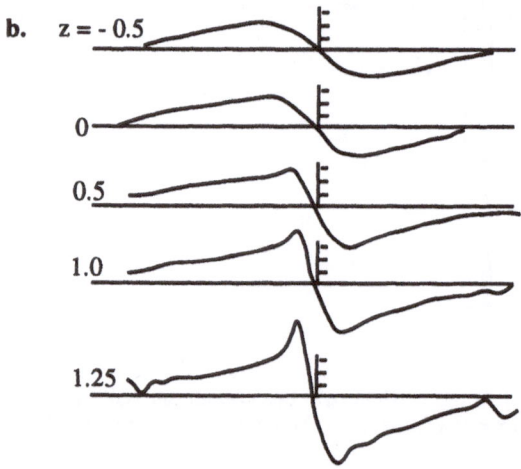

b. z = - 0.5

0

0.5

1.0

1.25

Fig. 7.10. Analytical continuation of the anomalous field of a perfectly conducting, vertical strip in conducting half-space. The field is normalized to the horizontal component of the magnetic field above the homogeneous half-space. **a.** In-phase and **b.** out-of phase components of the field. (Berdichevsky and Zhdanov 1984)

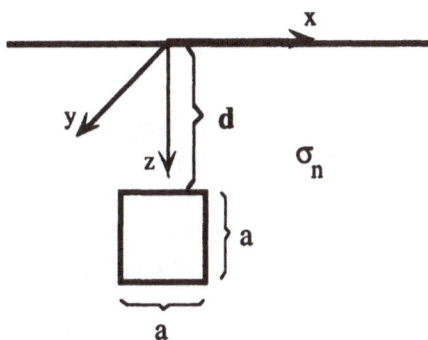

Fig. 7.11. The perfectly conducting, rectangular inclusion embedded in a conducting half-space. The inclusion is two-dimensional, **a/d** =1/2 and the electromagnetic wavelength in the half-space is 500 times d. (After Berdichevsky and Zhdanov 1984)

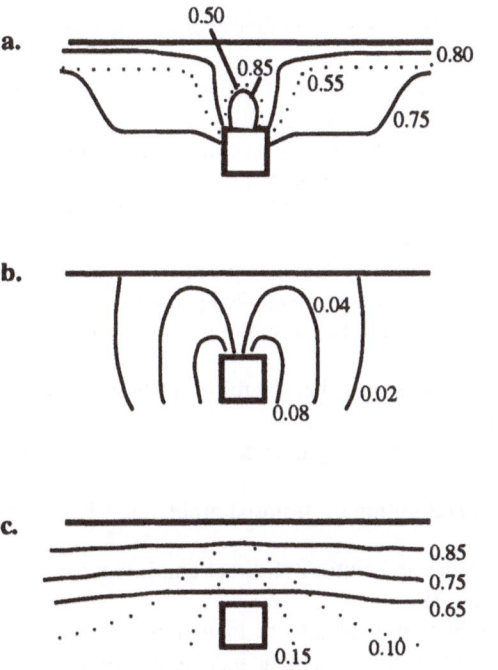

Fig. 7.12. Analytical continuation of the anomalous field of a conducting inclusion in conducting half-space. Each field is normalized to the horizontal component of the corresponding field above the homogeneous half-space. **Full lines:** contours of the real part; **dotted lines:** contours of the imaginary part. **a.** Horizontal magnetic (normalized) field; **b.** vertical magnetic (normalized) field. Only the real part is shown. The imaginary part almost coincides with the real curves; **c.** horizontal electric (normalized) field. (Berdichevsky and Zhdanov 1984)

7.3. MIGRATION

For wave fields, the continuation process, called migration has found wide practice in seismical prospecting (see Claerbout 1971, Aki and Richards 1980; Berkhout 1982, 1984, 1987, 1976; Hatton et al. 1986). The technique has been extended also to electromagnetic waves by Zhdanov and his coworkers (Berdichevsky and Zhdanov 1984, Zhdanov 1980, 1988 and references therein). Many practical algorithms of continuation and migration principles in fact involve also filtering, so that these principles can be thought of as a bridge between true inversion and data filtering.

REFERENCES

Aki K, Richards PG (1980) Quantitative seismology. Theory and methods, I & II. WH Freeman, San Fransisco

Berdichevsky MN, Zhdanov MS (1984) Advanced theory of deep geomagnetic sounding. Elsevier, Amsterdam, 408 pp

Berkhout AJ (1982) Seismic migration. Imaging of acoustic energy by wave field extrapolation. A.. Theoretical aspects. Elsevier, The Netherlands, 351 pp

Berkhout AJ (1984) Seismic migration. Imaging of acoustic energy by wave field extrapolation. B. Practical aspects. Elsevier, The Netherlands, 274 pp

Berkhout AJ (1987) Applied seismic wave theory. Elsevier, The Netherlands, 377 pp

Claerbout JF (1971) Toward a unified theory of reflector mapping. Geophysics 36: 467 - 481

Claerbout JF (1976) Fundamentals of geophysical data processing. McGraw-Hill, USA, 274 pp

Grant FS, West GF (1965) Interpretation theory in applied geophysics. McGraw-Hill, USA, 583 pp

Hatton L, Worthington MH, Makin J (1986) Seismic data processing: theory and practice. Blackwell London, 177 pp

Lavrentiev MM (1967) Some improperly posed problems of mathematical physics. Springer-Verlag, Berlin-Heidelberg, 72 pp

Oldenburg DW (1990) Inversion of electromagnetic data: an overview of new techniques. Surv in Geophys 11 (2-3): 231 - 270

Tikhonov AN, Arsenin VY (1977) Solution of ill-posed problems. VH Winston & Sons, Washington, DC, 258 pp

Varentsov IM (1985) Modern trends in solution of direct and inverse electromagnetic problems. Geophys Surv 6 (1-2): 55 - 78

Weidelt P (1972) The inverse problem of geomagnetic induction. Z Geophys 38: 257 - 289

Weidelt P (1981) Extremal models for electromagnetic induction in two-dimensional perfect conductors. J Geophys 49: 217 - 225

Weidelt P (1985) Construction of conductance bounds from magnetotelluric impedances. J Geophys 57: 191 - 206.

Zhdanov MS (1980) Cauchy integral analogues for the separation and continuation of electromagnetic fields within conducting matter. Geophys Surv 4: 115 - 136

Zhdanov MS (1988) Integral transforms in geophysics. Springer-Verlag, Berlin-Heidelberg, 367 pp

Zhdanov MS, Golubev NG (1983) Use of the finite functions method for the solution of the 2D inverse problem. Preprint 2a, IZMIRAN, Acad Sciences USSR, Moscow, 23 pp

Zhdanov MS, Varentsov IM (1983) Interpretation of local two-dimensional electromagnetic anomalies by formalized trial procedure. Geophys J R Astr Soc 75: 623 - 638

Zidarov D (1990) Inverse gravimetric problem in geoprospecting and geodesy. Elsevier & Publ House Bulg Acad Sci, Bulgaria, 283 pp

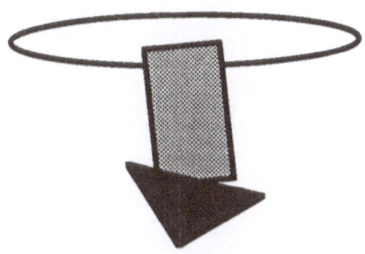

"Nothing is certain,
not even this statement"

(Multatuli, Dutch writer, 19th century)

CHAPTER 8

ERROR ANALYSIS

8.1 INTRODUCTION

Every quantitative inversion of measured data should contain some level of estimate on the resolution and accuracy of the parameters obtained. For analytically defined forward and inverse problems rather involved studies on the properties of the expressions, models and model parameters are possible. More often when working with real data, only an approximate estimation of the parameter errors can be performed. There is a wealth of variety to categorize errors of inversion. Hjelt (1973) has discussed the problem of errors in geophysical interpretation suggesting the division of errors given in Fig. 8.1

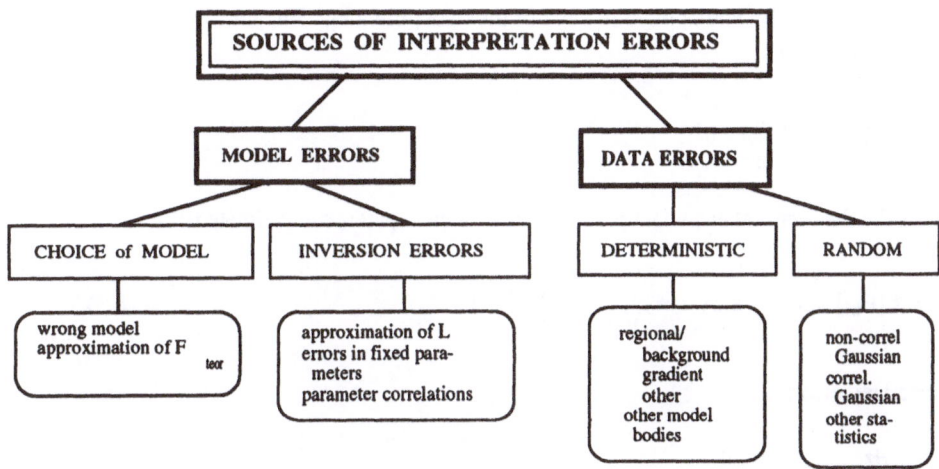

Fig. 8.1. Classification of errors in (geophysical) inversion. (after Hjelt 1973, 1974)

The original papers use disturbing fields instead of data errors and some other minor modification have been made in redrawing the scheme. The interdivision can be discussed and arguments can be presented for additional sources, but the scheme concentrates on the most obvious ones.

8.2 ON LINEARIZED ERRORS

One may perhaps say that a great majority of inversion procedures end up with a linearization, if not during the last refinement step of optimization. Thus, it is no great surprise that the majority of error analysis is also based on linearized theory. Some care is, however, required since for a strongly non-linear parameter dependence one soon exceeds the boundary of validity for the linearized approach. This has been considered in the context of confidence intervals for two-parameter cases as by Johansen (1977), Pedersen (1979) and Jakobsen (1987). It is instructive enough to use our basic function as a tool to demonstrate the difference between complete error and linearized error.

Consider the field of a line source, located at point P_0 (x_0, z). It produces the field (data) vector:

$$F_h = \frac{2m \cdot u}{z^2 + u^2} \qquad\qquad \text{with } u = x - x_0 \qquad\qquad (8.1)$$

Let the only unkown parameter be the non-linear depth to the line source, z. The inverse operator can be determined analytically and applied to each of the data points F_{hi}. We get as many estimates as there are data points (N):

$$z_i = L\{F_{hi}\} = \sqrt{\frac{2m \cdot u_i}{F_{hi}} - u_i^2} \ . \qquad\qquad (8.2)$$

If F_{ei} is the data error at each data point the error of each parameter estimate is:

$$\Delta z_{ex} = L\{F_{hi}\} - z_{true} = z_{ex} \cdot (-1 \pm \sqrt{1 - \frac{2m \cdot u_i F_{ei}}{F_{hi}^2 \cdot z_{ex}^2}}) \ , \qquad\qquad (8.3)$$

and the linearized error:

$$\Delta z_{lin} = \frac{\partial z}{\partial F_{ti}} \cdot F_{ei} = -\frac{m \cdot u_i F_{ei}}{F_{hi}^2 \cdot z_{ex}} \ . \qquad\qquad (8.4)$$

The latter expression follows naturally also from the complete error Eq. (8.3) by expanding it into a series and leaving only the first-order term. In Table 8.1 some numerical values are given. Only the right half of the profile is considered, since the anomaly is of minimum-maximum type and hence antisymmetric with respect to u = 0 ($x_i = x_0$). The relative errors are smallest for data points around the anomaly maximum, where also the difference between the complete and linearized error is small. Far from the maximum, the relative error increases rapidly and so does the difference between the error estimates. The result, although self-evident, is a guide for considering the error estimates in higher dimensions, where visualization is more difficult. Close to the half-value point, the difference is already 40%, when $F_{ei} = -\text{const.} = 0.1$.

TABLE 8.1   Basic function Eq (8.1), its complete and linearized errors at selected points.

	0.5	1	2	3	4
u_j	0.5	1	2	3	4
F_{hi}	0.800	1.001	0.800	0.600	0.470
$\Delta z_{compl} / z_{ex}$	0.085	0.106	0.310	0.731	1.370
$\Delta z_{lin} / z_{ex}$	0.063	0.100	0.312	0.834	1.810

8.3 MINIMAL ERROR IN LEAST-SQUARES INVERSION

When all deterministic components of the error between data and model fields have been eliminated, the remainder is a pure random function. Statistical estimation theory states that there exists a smallest parameter variance (see Rao 1965; Whalen 1971). This smallest value is called the Cramer-Rao bound of the parameter. It occurs when the errors e are independent of each other at the various measuring points and in addition follow Gaussian statistics. For $e = N(0, \sigma)$ the Cramer-Rao bound is:

$$\Delta p_j = \sigma \sqrt{A_{jj}}, \qquad\qquad j = 1, 2, ..., M \qquad\qquad (8.5)$$

$$\text{where } \sigma = \sqrt{\frac{\sum_{i}^{N} e_i^2}{N - M}}$$

is the mean variance of the inversion, M the number of parameters, N the number of observations and **A** the covariance matrix, the inverse of the information matrix **T** of the inverse model.

$$A = T^{-1} \qquad\qquad (8.6)$$

$$T = \{t_{jk}\}, \quad t_{jk} = \sum_{i}^{N} \left(\frac{\partial F_{ti}}{\partial p_j} \cdot \frac{\partial F_{ti}}{\partial p_k} \right) \qquad j, k = 1, 2,...., M.$$

It can be shown that the minimum of the Cramer-Rao bound of each parameter variance is reached when the corresponding parameter is inverted alone. In other words:

$$\Delta p_j \geq \Delta p_{j, min} = \sigma / \sqrt{t_{jj}} \qquad\qquad (8.7)$$

will be the smallest obtainable parameter error in least-squares inversion. This minimum parameter error depends among other things on the separation of data points and the length of the interpretation profile. The influence of these factors on minimum parameter errors of magnetic inversion using the thick plate model was studied in Hjelt (1974). Examples of this analysis are shown in the following.

In the Figs. 8.2 and 8.3, the variation of the minimum parameter error is shown, when the centre, XK, of the data profile and thus the whole profile is varied with respect to the plate. The anomalies have been calculated for the inclination of the Earth's geomagnetic field I = 75°, typical for northern Finland and Scandinavia. The plate dip is assumed also to be 75°, so that the plate is magnetized along the dip. The anomalies are symmetric with respect to the centre of the plate and so are minimum error curves for width, horizontal location and depth to the top of the plate accordingly symmetric with respect to XK = 0. The curves confirm the intuitively natural result that there exists a region, about the size of the profile length, around the centre of the plate, where minimum errors for these three parameters are obtained.

In the second group of minimum error curves, for dip, susceptibility and depth extent of the plate, the maxima of the anomaly profiles change position with dip. The location of the maxima are indicated by vertical arrows in each of the three subsequent parts of Fig. 8.3. The minimum error curves of susceptibility are centred close to (but not exactly at) the maxima. They do not depict a similar flatness than the preceding curves. This is an indication of how closely the susceptibility parameter is tied to the maximum amplitude of the anomaly profile. The error curves for dip has a complicated behaviour, the minimum error occurring closer to the steep flanks of the anomaly than to the maximum itself. The choice of data points on a profile to be used for minimum error dip determination has to be made more carefully than for the previous parameters.

In earlier discussion on the properties of magnetized plate models (e.g. Chap. 4) it was evident that the depth extent (h) and the dip of a plate were strongly correlated and that the depth extent affects the flanks of the anomaly far more than the maximum. It is thus no surprise, that the behaviour of the minimum error curves for depth extent are also strongly asymmetric with respect to the anomaly maximum. A closer examination of the curves discloses that the smallest errors of h occur when the data profile is located above the lower horizontal surface of the plate (shown by the dots in the models of the corresponding figure.

8.4 CORRELATION BETWEEN PARAMETERS

Because of the definition of the Cramer-Rao bounds, the correlation between model parameters will affect essentially the amount of parameter error in geophysical inversion. An early analysis of correlation coefficients between the parameters of a magnetized, vertical thick plate has been presented by Kalinina and Holzmann (1967, 1971). The minimum parameter error bounds stand out clearly among the various combinations of parameters used in inversion (Table 8.2).The latter paper contains also a study of the influence of a priori parameter boundaries on the accuracy of parameter determination. Some characteristic results are presented in Table 8.3 and demonstrate the positive effects of constraining the parameters, let it be based on other geophysical models, intuition or some other source. The correlations between selected parameter combinations are shown in Table 8.4.

a.

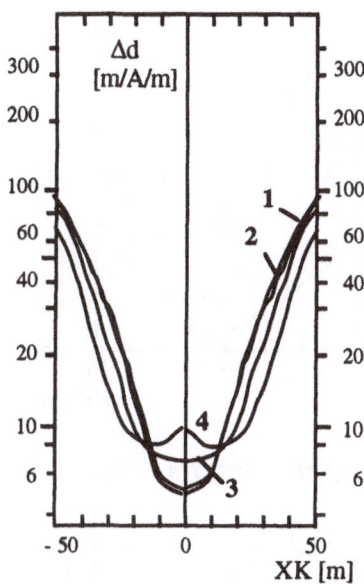

b.

c.

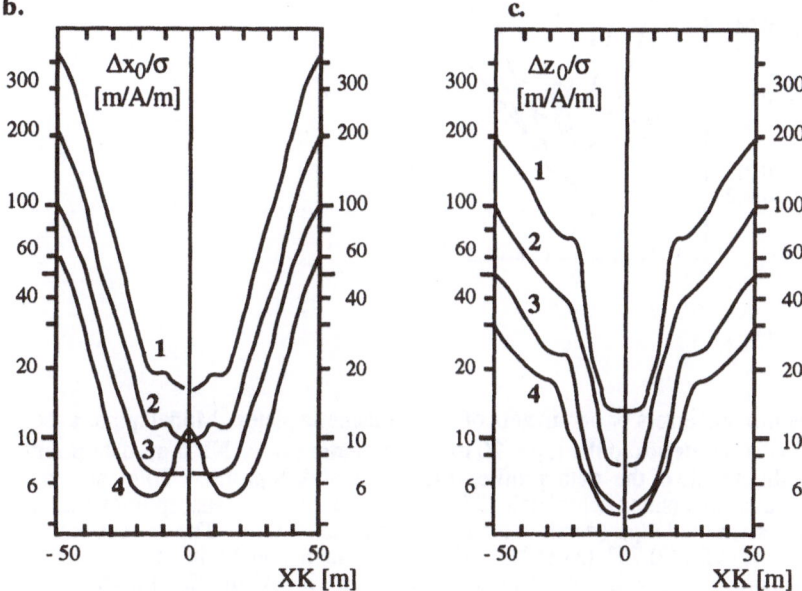

Fig. 8.2. Minimum errors or parameters of thick magnetic plates. Model parameters: **depth** to the top = 10 m, **depth extent** = ∞, **dip** = 75°, **varying thickness d** = **(1)** 5 m; **(2)** 10 m; **(3)** 20 m; **(4)** 30 m. **NP** # of data points = 11. **XK** the middle of the data profile, **XL** length of data profile = 20 m and σ = mean variance of inversion (After Hjelt 1974). **a.** Width of the plate; **b.** horizontal position of the plate; **c.** depth to the top of the plate

a.

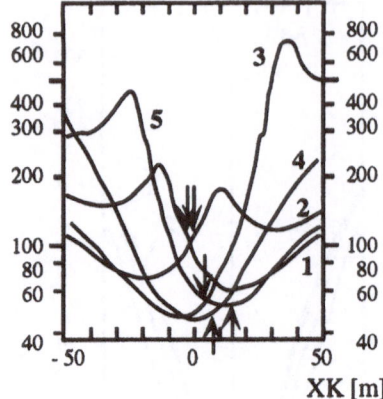

b.

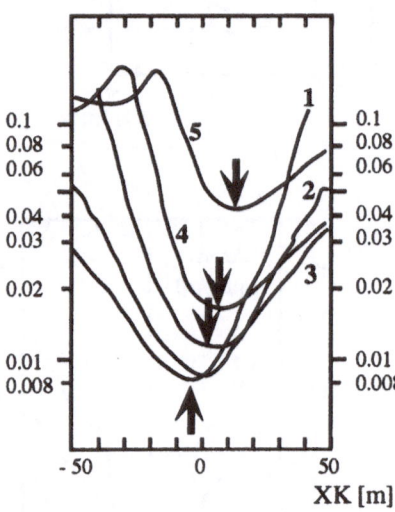

Fig. 8.3. Minimum errors or parameters of thick magnetic plates. Model parameters: thickness **d** = 10 m; **depth** to the top = 20 m; **depth extent** = ∞. **NP** # of data points = 11. **XK** the middle of the data profile, **XL** length of data profile = 20 m and σ = mean variance of inversion (After Hjelt 1974). **a.** Dip; **b.** magnetic susceptibility; **c.** varying depth extent (and dip) (depth to the top of the plate = 10 m). The curves correspond to dip = (**1**) 95°, (**2**) 75°, (**3**) 55°, (**4**) 35°, (**5**) 15° in **a** and **b**; in **c** the parameter combinations are: (**1**) dip = 35°, h = 100 m, (**2**) dip =75°, h = 50 m, (**3**) dip =55°, h = 50 m, (**4**) dip =35°, h = 50 m

(Fig. 8.3 continues)

(Fig. 8.3 continued)

c.

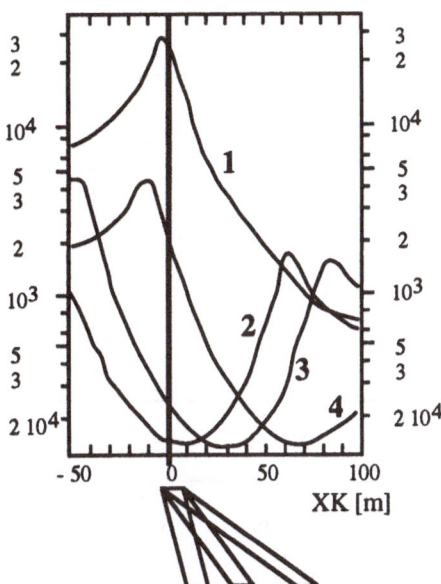

The Tables have been presented here for reasons of simplicity so that the plate parameters have been numbered from 1 to 6 (cf. the Table captions) in the same order as when defining the plate parameters in Chapter 4. Kalinina and Holzmann fixed the dip of the plate (vertical) and their group of parameters contains the strike direction instead. It has been given the parameter # of the dip in the tables. In Table 8.2 the first two columns give the minimum error (the Cramer-Rao bound, when a single parameter is inverted) and the maximum error, when all six plate parameters were inverted. The sensitivity of the parameters varies strongly as the third column on the quotient (not originally presented by Kalinina and Holzmann) between these extremal values shows. By studying the following columns (where the included parameters are indicated by numbers above each column) gives an idea of which parameter combinations are problematic and which parameters should be preferably be fixed or constrained as much as possible.

It is evident without doubt that the parameter combination 346 (z_o, h and J) is very harmful for the accuracy of parameter determination. Constraining even one of these parameters (Table 8.3) reduces their errors by an order of magnitude. In order to reduce the errors of parameters 1 and 2 (d and x_0) either or both have to be constrained. In their analysis, Kalinina and Holzmann assumed the constraints of the parameters to be additive to the normalized covariance matrix $\mathbf{T}^{-1}$ and assumed that each constraint was given by $2 \cdot (\Delta p_i)^2$:

$$\underline{\mathbf{B}} = \underline{\mathbf{T}}^{-1} + \underline{\mathbf{S}}^{-1} \qquad \text{with } [\underline{\mathbf{S}}^{-1}]_{ii} = 2 \cdot (\Delta p_i)^2. \qquad (8.8)$$

TABLE 8.2 Parameter errors , N = 21 points/profile, L = 5 profiles, $\Delta x = 0.2$, $\Delta y = 0.5$ (Kalinina and Holzmann, 1967). Plate parameters are numbered 1 - 6 as follows: 1 = d, 2 = x_0, 3 = z_0, 4 = h, 5 = α (=strike!!) and 6 = J (magnetization, not susceptibility)

	min	max	max/min
Δd	0.158	0.546	3.5
Δx_0	0.130	0.226	1.8
Δz_0	0.137	4.61	34
Δh	0.186	13.75	74
ΔJ	0.126	12.94	102

	12345	13456	12456	12356	12346	23456	1234	1236	1246	1346	2346
Δd	0.534	0.546	0.527	0.534	0.546	-	0.534	0.540	0.527	0.546	-
Δx_0	0.130	-	0.226	0.225	0.130	0.226	0.130	0.130	0.130	-	0.130
Δz_0	0.345	4.61	-	0.346	4.61	4.45	0.345	0.627	-	4.61	4.45
Δh	0.896	13.75	1.87	-	13.75	13.61	0.896	-	1.870	13.75	13.60
ΔJ	-	12.94	0.971	0.126	12.94	12.66	-	0.187	0.971	12.90	12.66

	123	124	134	136	146	234	236	246	346
Δd	0.182	0.348	0.597	0.540	0.527	-	-	-	-
Δx_0	0.130	0.130	-	-	-	0.130	0.130	0.130	-
Δz_0	0.158	-	0.387	0.627	-	0.225	0.311	-	4.45
Δh	-	0.411	1.000	-	1.870	0.306	-	0.950	13.61
ΔJ	-	-	-	0.843	0.971	-	0.285	0.642	12.66

	12	13	14	16	23	24	26	34	36	46
Δd	0.158	0.182	0.348	0.267	-	-	-	-	-	-
Δx_0	0.130	-	-	-	0.130	-	0.130	-	-	-
Δz_0	-	0.158	-	-	0.137	-	-	0.225	0.310	-
Δh	-	-	0.411	-	-	0.206	-	0.306	-	0.340
ΔJ	-	-	-	0.213	-	-	0.126	-	0.284	0.392

TABLE 8.3. Parameter errors when a priori parameter bounds are specified (from Kalinina and Holzmann, 1971). Plate parameters are numbered 1 - 6 as follows: $1 = d$, $2 = x_0$, $3 = z_0$, $4 = h$, $5 = \alpha$ (=strike, not dip!!) and $6 = J$ (magnetization, not susceptibility)

	1 $2(\Delta p_j)^2$	δp_j	2 $2(\Delta p_j)^2$	δp_j	3 $2(\Delta p_j)^2$	δp_j	4 $2(\Delta p_j)^2$	δp_j	5 $2(\Delta p_j)^2$	δp_j	6 $2(\Delta p_j)^2$	δp_j
Δd	∞	0.546	∞	0.528	∞	0.540	∞	0.534	1	0.426	1	0.407
Δx_0	∞	0.226	∞	0.226	∞	0.225	∞	0.226	∞	0.131	0.1	0.121
Δz_0	∞	4.61	0.5	0.698	∞	0.708	∞	0.427	∞	0.477	1	0.425
Δh	∞	13.75	∞	2.79	1	0.997	∞	1.170	1	0.794	1	0.798
ΔJ	∞	12.98	∞	2.18	∞	1.260	0.5	0.706	1	0.772	1	0.720
$\Delta \alpha$	∞	0.184	∞	0.184	∞	0.184	∞	0.184	∞	0.004	10^{-4}	0.0099

TABLE 8.4. Selected correlation coefficients for same priori parameter bounds as in Table 8.2. [Columns are numbered as in Table 8.2] (Kalinina and Holzmann 1971). Plate parameters are numbered 1 - 6 as follows: $1 = d$, $2 = x_0$, $3 = z_0$, $4 = h$, $5 = \alpha$ (= strike, not dip!!) and $6 = J$ (magnetization, not susceptibility)

	1 r_{ij}	2 r_{ij}	3 r_{ij}	4 r_{ij}	5 r_{ij}	6 r_{ij}
r_{12}	0.0015	0.0016	0.0015	0.0015	- 0.018	*
r_{13}	- 0.259	- 0.041	- 0.773	- 0.620	- 0.657	- 0.613
r_{14}	0.143	- 0.547	0.0105	- 0.714	- 0.361	- 0.399
r_{16}	- 0.204	0.297	- 0.638	- 0.011	- 0.434	- 0.362
r_{23}	*	*	0.0017	0.0021	0.021	*
r_{24}	*	0.0012	*	*	*	*
r_{26}	*	- 0.0011	*	*	0.018	*
r_{34}	- 0.991	- 0.741	- 0.467	- 0.176	- 0.137	- 0.128
r_{36}	0.997	0.896	0.920	0.587	0.821	0.790
r_{46}	- 0.998	- 0.955	- 0.743	- 0.641	- 0.613	- 0.641

$* = < |0.001|$

The selected coefficients confirms the conclusion stated above about the interplay between parameters 3, 4 and 6. Without constraints the correlation coefficients r_{34}, r_{36}, and r_{46} are very close to 1 in absolute value. The constraints in columns 4 to 6 shift part of the correlation load to the interplay between parameters 1 and 3 and 1 and 4. This explains why reducing the accuracy of parameter 1 is difficult.

8.5 ADVANCED CONCEPTS OF ERROR ANALYSIS

Van der Sluis and van der Vorst (1987) have discussed the solution of the linear parameter problem, especially considering the sparse matrices in seismic tomography. They propose to apply a regularization procedure in order to circumvent the dramatic effects of small eigenvalues of the multiplying matrix. They present an error analysis related to the standard regularization procedure.

Instead of solving:

$$\underline{G}\, p = F_h,$$

the least squares problem:

$$\begin{bmatrix} \underline{G} \\ \kappa \underline{I} \end{bmatrix} p = \begin{bmatrix} F_h \\ 0 \end{bmatrix} \tag{8.9}$$

is solved. Using the singular value decomposition described earlier, the solution can be expressed as:

$$p = \underline{V}\,(\Delta^T \Delta + \kappa I)^{-1} \Delta\, \underline{U}^T F_h, \tag{8.10}$$

and the parameter error as:

$$\Delta p = p - p^* = \Delta\varepsilon_{appr} + \Delta\varepsilon_{data} \tag{8.11}$$

with $\Delta\varepsilon_{appr} = \underline{V}\,(\Phi - \Delta^{-1})\underline{U}^T \underline{G}\, p^*,$

$$\Delta\varepsilon_{data} = \underline{V}\,\Phi\,\underline{U}^T \varepsilon \quad \text{and}$$

$$\underline{\Phi} = \text{diag}\,[\phi\,(\lambda_j)] \qquad\qquad \phi(\lambda_j) = \frac{\lambda_j}{\lambda_j^2 + \kappa^2}.$$

If the data are uncorrelated with zero average and equal variance σ^2, then:

$$\|\Delta\varepsilon_{appr}\|^2 = \kappa^4 \sum_j \left[\frac{(\underline{U}^T F_h)_j}{(\lambda_j^2 + \kappa^2)\,\lambda_j} \right]^2 \quad \text{and} \tag{8.12 a}$$

$$E\left[\|\Delta\varepsilon_{data}\|^2\right] = \sigma^2 \sum_j \left[\frac{\lambda_j}{\lambda_j^2 + \kappa^2} \right]^2. \tag{8.13 a}$$

The solution is thus biased since $\Delta p_{appr} \neq 0$. At the same time the variance has been reduced and the trade-off between the bias and the variance is controlled by the choice of κ.

Van der Sluis and van der Vorst (1987) also consider another regularization procedure: to substitute 0 for $(\underline{U}^T F_h)_j$ when the jth eigenvalue λ_j is below a threshold η. Then:

$$\|\Delta\varepsilon_{appr}\|^2 = \sum_{\lambda_j < \eta} \left[\frac{(\underline{U}^T F_h)_j}{\lambda_j}\right]^2 \quad \text{and} \tag{8.12 b}$$

$$E\left[\|\Delta\varepsilon_{data}\|^2\right] = \sigma^2 \sum_{\lambda_j \geq \eta} \left[\frac{1}{\lambda_j}\right]^2. \tag{8.13 b}$$

Wunsch (1990) discusses the overdetermined linear problem:

$$\underline{G}\, p = F_h.$$

If the noise in the data has a zero mean and covariance $\underline{R}$, then the solution is:

$$p = (\underline{G}^T \underline{R}^{-1} \underline{G})^{-1} \underline{G}^T \underline{R}^{-1} F_h = \underline{G}^+ F_h, \tag{8.14}$$

where $\underline{G}^+$ is the generalized inverse of the original equation system. The error covariance of the solution is

$$E(p^2) = \underline{G}^+ \underline{R}\, \underline{G}^{+T} = (\underline{G}^T \underline{R}^{-1} \underline{G})^{-1}. \tag{8.15}$$

In the case where the covariance matrix has the familiar form $\underline{R} = \sigma^2 \underline{I}$ as discussed, then the error covariance reduces to:

$$E(p^2) = \sigma^2 (\underline{G}^T \underline{G})^{-1}.$$

If one, in addition to the equation specifies the a priori solution covariance:

$$\underline{S} = <p\, p^T> \tag{8.16}$$

a useful solution to the original equation system is, according to Wunsch:

$$p = \underline{S}\, \underline{G}^T (\underline{G}^T \underline{S}\underline{G} + \underline{R})^{-1} F_h =$$

$$= (\underline{G}^T \underline{R}^{-1} \underline{G} + \underline{S}^{-1})^{-1} \underline{G}^T \underline{R}^{-1} F_h. \tag{8.17}$$

The corresponding error covariances are:

$$E(p^2) = \underline{S} - \underline{S}\,\underline{G}^T\,(\underline{G}\,\underline{S}\,\underline{G}^T + \underline{R})^{-1}\,\underline{G}\,\underline{S}$$

$$= (\underline{S}^{-1} + \underline{G}^T\underline{R}^{-1}\underline{G})^{-1}. \tag{8.18}$$

Suppose now that new observations F_{h2} are made with the error covariance R_2 independent of the first data set. The amount of new data keeps the original equation still well overdetermined. A solution can be written without inverting the new, larger matrix. Instead the existing solution to p (in the form of the estimates given above) is used. The "recursive weighted least squares" (Wunsch 1990) minimizes the weighted residuals of both the new and the old set of equations:

$$e = \begin{pmatrix} e_1 \\ e_2 \end{pmatrix}^T \begin{pmatrix} R_1^{-1} & 0 \\ 0 & R_2^{-1} \end{pmatrix} \begin{pmatrix} e_1 \\ e_2 \end{pmatrix} \qquad \text{with} \quad e_i = F_{hi} - \underline{G}_{1i}\,p \quad i = 1, 2. \tag{8.19}$$

The solution is

$$p_2 = [\underline{G}_1^T\,R_1^{-1}\,\underline{G}_1 + \underline{G}_2^T\,R_2^{-1}\,\underline{G}_2]^{-1}\,[\underline{G}_1^T\,R_1^{-1}\,F_{h1} + \underline{G}_2^T\,R_2^{-1}\,F_{h2}], \tag{8.20 a}$$

which can "easily" be manipulated into a simpler form:

$$p_2 = p_1 + K_2\,[F_{h2} - \underline{G}_2\,F_{h1}], \tag{8.20 b}$$

with

$$K_2 = E(p_1^2)\,\underline{G}_2^T\,[\underline{G}_2\,E(p_1^2)\,\underline{G}_2^T + R_2]^{-1}. \tag{8.21 a}$$

The second term of the solution is the difference between the new data and the "prediction" of these data based upon the previous estimate of the parameter vector.

The covariance of this improved parameter estimate is:

$$E(p_2^2) = E(p_1^2) - K_2\,\underline{G}_2\,E(p_1^2), \tag{8.22 a}$$

which can be studied qualitatively following Wunsch (1990). If the initial error estimate is diagonal:

$$E(p_1^2) = \Delta^2 I \tag{8.22 b}$$

then $\quad K_2 = \underline{G}_2^T\,(\underline{G}_2\,\underline{G}_2^T + R_2/\Delta^2)^{-1}. \tag{8.21 b}$

If the norm of the covariance matrix of the new data is small compared with the initial error estimate. This means that the new set of data themselves form an underdetermined system. Then the second term in the last expression is negligible and:

$$K_2 \rightarrow \underline{G}_2^+ = \underline{G}_2^T\,(\underline{G}_2\,\underline{G}_2^T)^{-1} \tag{8.21 c}$$

and

$$\mathbf{p}_2 \rightarrow (\mathbf{I} - \underline{G}_2^+ \underline{G}_2) \, \mathbf{p}_1 + \underline{G}_2^+ \, \mathbf{F}_{h2} \,. \tag{8.20 c}$$

The generalized inverse matrix $\underline{G}_2^+$ is the same as for an underdetermined system. Should the new observations be fully determined, the new, low noise data produce a solution to replace the previous estimate completely. If the new data are not able to produce estimates for all elements of $\mathbf{p}$, then those components of the old estimate lying in the null-space of the new observations will be retained.

If the new data are very noisy, then the matrix $\mathbf{K}_2$ produces no or only minor changes to the previous estimate. In the general case, the new solution is a properly weighted average of the estimates obtained from the previous and the new data.

REFERENCES

Hjelt S-E (1973) On the error analysis of interpretatation in applied geophysics. In: Jatila E (ed): Natl Conf on Geophysics, Helsinki, 29.-30.5.1973, Helsinki, pp 89 - 98 (in Finnish)

Hjelt S-E (1974) A contribution to the quantitative error analysis in magnetic and gravimetric interpretation. Geophys Prosp 22: 546 - 567

Jacobsen HB (1987) The radius-ratio test in nonlinear geophysical modelling. In: Vogel A (ed): Model optimization in exploration geophysics. Proc 4th Int Mathematical Geophysics Seminar, Freie Univ Berlin, Feb 6 - 8, 1986. Fr Vieweg and Sohn. Braunschweig/ Wiesbaden, pp 373 - 396

Johansen H-K (1977) A man/computer interpretation system for resistivity soundings over a horizontally stratified earth. Geophys Prosp 25: 667 - 691

Kalinina TB, Holzmann (Golt'sman) FM (1967) A statistical interpretation algorithm and its application in the solution of the inverse problen of magnetic exploration. Izv Akad Nauk SSSR, Phys Solid Earth, (7): 451 - 457

Kalinina TB, Holzmann (Golt'sman) FM (1971) Analytical comparison of methods of geophysical interpretation. Izv Akad Nauk SSSR, Phys Solid Earth, (5): 320 - 325

Pedersen LB (1979) Constrained inversion of potential field data. Geophys Prosp 27: 726 - 748

Rao CR (1965) Linear statistical inference and its applications. John Wiley and Sons, New York, 522 pp

van der Sluis A, van der Vorst HA (1987) Numerical solution of large, sparse linear algebraic systems arising from tomographic problems. In: Nolet G (ed): Seismic tomography. (With applications in global seismology and exploration geophysics). D. Reidel, Dordrecht, pp 49 - 83

Wunsch C (1990) Using data with models; ill-posed problems. In: Desaubies Y, Tarantola A, Zinn-Justin J (eds): Oceanographic and geophysical tomography. NATO Adv Study Inst, Session L, 9.8. - 3.9. 1988. North-Holland, Amsterdam, pp 203 - 248

Chapter 9

PARALLEL COMPUTATION in MODELLING and INVERSION

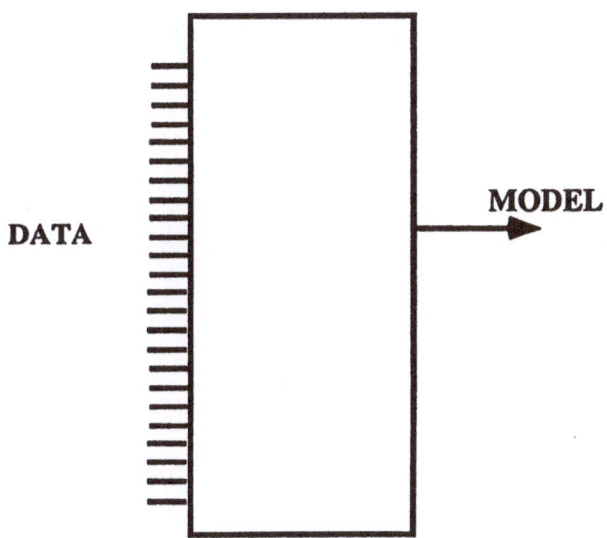

CHAPTER 9

PARALLEL COMPUTATION in MODELLING and INVERSION

The advances in computer technology has brought a new tool for solving computationally intensive problems, parallel computer and parallel computation. This tool has found widespread use in geophysical, both forward (direct) and inverse solutions. Most approaches are very straightforward in that they are used only to solve large systems of equations. Parallel processing and inversion algorithms have been rather thoroughly covered for seismic methods (e.g. the excellent collection of articles, edited by Eisner 1989; Mora 1990). Less work has been done with regard to other geophysical methods. A closer study reveals that there are several levels of inherent parallelism in most computations related to geophysical inversion (Hjelt 1989). These levels are briefly described in the following as well as some suggested solutions of implementing parallel computation for the benefit of geophysical inversion.

When discussing parallel computation one has to define:
1. The types of parallelism:
    * program = simultaneous tasks;
    * geometrical = relating to data structure, e.g. in numerical solution of differential equations;
    * algorithmic.
2. The architecture of the computing system:
    * SISD = singular instruction, singular data (= traditional single-processor sequential computation);
    * MISD = multiple instruction, singular data;
    * SIMD = single instruction, multiple data
    * MIMD = multiple instruction, multiple data.
3. The topology of the processors:
    * ring;
    * star;
    * network;
    * systolic array;
    * hypercube.

To gain the highest advantage of parallelism, the algorithmic approach, MIMD architecture and the hypercube topology seem to be most important. The algorithmic approach requires that the whole computation process needs to be reconsidered from the very roots of any problem. A very recent example demonstrating the amount of necessary rethinking has been described for seismic modelling by Myczkowski et al. (1991). A massively parallel machine with more than 65 000 processors was employed for this work, resulting in impressive calculation rates in excess of 15 Gflops (= $15 \cdot 10^9$ floating point operations per second).

In the following more modest aspects of parallelism are discussed. Some key points are highlighted especially from the viewpoint of geophysical inversion. For a general background on various aspects of parallelism a wealth of literature is fortunately available, e.g. Hockney and Jesshope (1981), Miranker (1971) on numerical analysis, Hwang and Briggs (1986), Karp (1987), Ortega and Voigt (1985) on differential equations and White (1987) on iterative methods. Architectural and topological questions are discussed in e.g. Hayes et al. (1986), Kung et al. (1987) and Hillis (1987).

9.1 Parallelism in Geophysical Problems

Parallel computation in seismology have recently been well covered as in the articles of Baker, Moorehead and by Mora and Tarantola in the collection by Eisner (1989) and by Mora (in Desaubies et al. 1990). The obvious parallel aspects of geophysical modelling and inversion for any method are easily recognized (Hjelt 1989). Alanko (1989) defined the main levels of parallelism in geophysical inversion to be parallelism in computation, when the model field is calculated at N points simultaneously and parallelism in data when K profiles are inverted simultaneously. A more detailed analysis leads to a further subdivision into (Hjelt 1989, Alanko 1989):

1. Comparison of various geophysical measurements ("complex interpretation") (Fig. 9.1);
2. Spectral processing (very typical in magnetotelluric data processing);
3. Numerical modelling (e.g. finite difference and element techniques);
4. Optimization of profile calculation by separating field point-dependent and other elements of the equations (cf. Hjelt 1973 for a discussion on the dipping plate case);
5. Non-linear optimization as such ; and
6. Subdivision of long profile in partial anomalies during inversion (Fig. 9.2), which all contain significant parts expedient to parallel computation.

An early computer program system for magnetic profile interpretation made use of several aspects of parallelism (Hjelt 1973). In a complete geophysical interpretation system parallel and serial parts form a complicated network of operations.

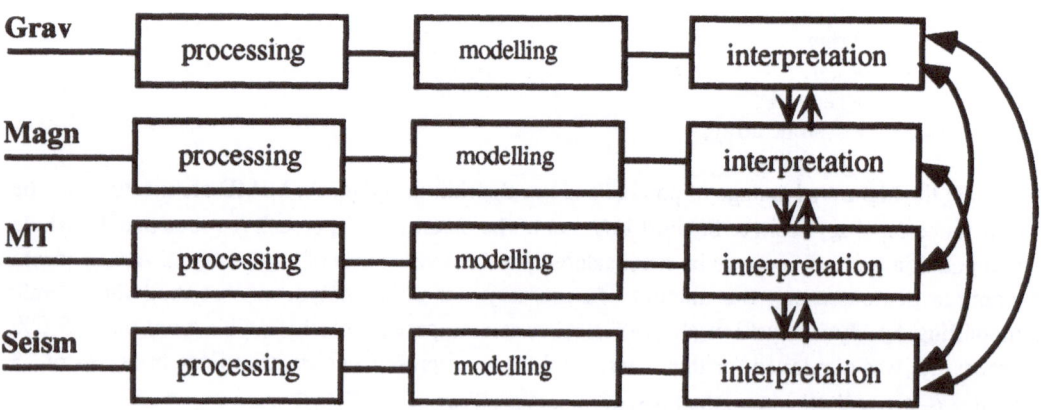

Fig. 9.1. Parallelism in geophysical "complex" interpretation: data from different methods are processed and modelled separately, but interpreted jointly in parallel mode (model comparison)

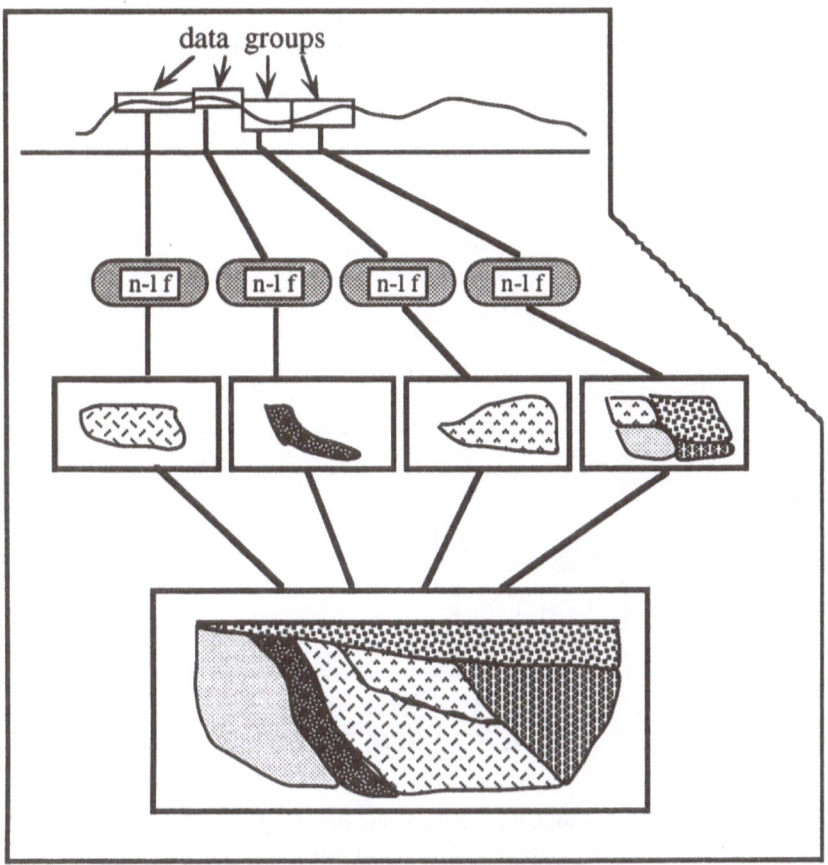

Fig. 9.2. The principle of partial anomalies in profile interpretation: each part is inverted separately using the same or even different inversion filters (**n-l f**). The filters are non-linear in the general case. The principle is especially appropriate for parallel computer implementation

9.2   FORWARD PROBLEMS

Much research and development has been made to develop efficient techniques and algorithms for calculation of the theoretical anomalies, i.e. for solving the forward problem. The most computer intensive solutions are needed for seismic and for electromagnetic modelling. Parallelism can be used to a great advantage in speeding up the forward problem by clever division of the problem. One simple idea is demonstrated in Fig. 9.3. Instead of performing the same calculation sequentially in one processor, N processors are employed simultaneously, each of them performing the same mathematical operations. Thus the time for solving the anomaly field f(x) along a profile with NX points, only a computer time proportional to NX/N is required. It is self-evident that the benefit increases with the number of processors available and when the length of the profile (or the region covered by profiles) increases. The principle of separating between operations depending on the field point and those independent thereof (as was described in Chapter 1 in connection with the magnetic plate model) will accelerate the calculation further.

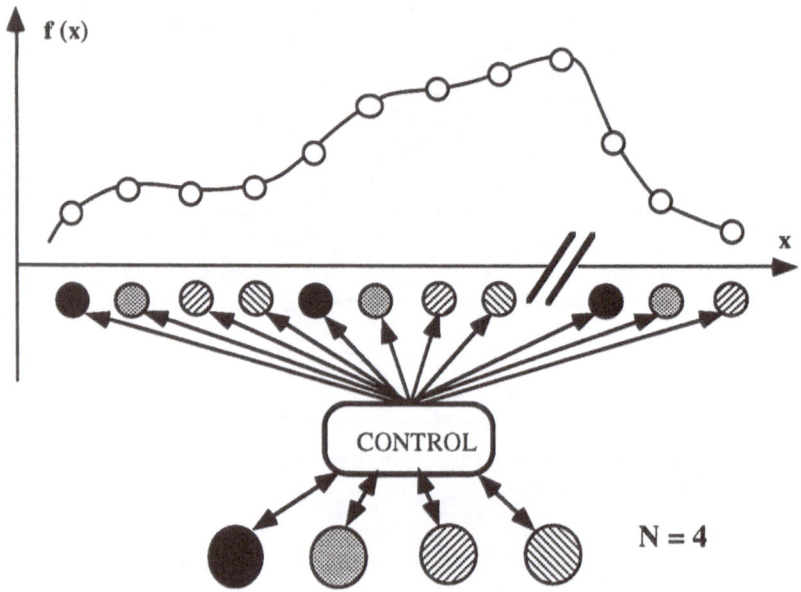

Fig. 9.3. Parallelism in anomaly computation: calculation of a geophysical profile is distributed among N processors. The patterns below each profile point identify the processor partipating in producing the calculated field at that point. A control processor or a procedure in the central processor takes care of bookkeeping = timing of the data flow between processors and points of storage of the results and keeping the processor busy in an optimal way (After Alanko 1989)

9.3 INVERSION BY OPTIMIZATION

Alanko (1989) has described experiments with a two-processor (later four processor) system in magnetic profile inversion using the two-dimensional plate model. The experimental system consists (cf. Fig. 9.4) of an Apple Macintosh IIx workstation, 4 LEVCO transputers, Think C (Symantec Corp), Parallel C (Logical Systems), MPW/Fortran and Occam 2 programming languages. The Cricket Graph application program for graphical output is used during the experimental stage to minimize additional programming. The communication between the transputers and the geophysical subprograms was designed in transputer-C. A full parallel operating system will reduce the amount of programming, but leads to problems of memory allocation: a large operating system occupies a significant part of the memories of the processors.

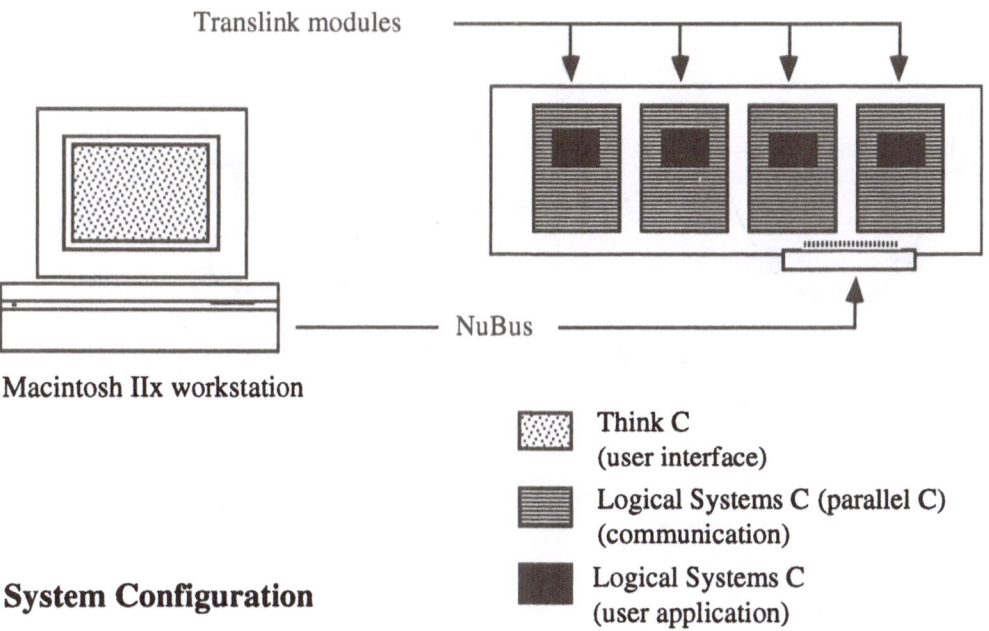

Translink modules

NuBus

Macintosh IIx workstation

System Configuration

▨ Think C
(user interface)

▤ Logical Systems C (parallel C)
(communication)

▪ Logical Systems C
(user application)

Fig. 9.4. Main components of the parallel microcomputer system used in the study of Alanko (1989)

The parameters of the plates were found by non-linear optimization and the principle of partial profiles (Figs. 9.5 and 9.6) was applied. The results were rather straightforward and the number of processors was too small to obtain great saving of computer time in these experiments. More encouraging results were obtained when the subsurface is covered with plates of the same size and form and the unknown linear parameters, the magnetic susceptibility of the plates, were recovered by the iterative parallel Gauss-Seidel method (Fig. 9.7).

The speedup caused by parallelization will depend on several factors, such as the size of the problem (number of unknown parameters), number of available processors and the structure of the algorithm with respect to parallelization. Almost complete (>95%) speedup was obtained for the linear magnetometric inversion with the experimental two-transputer system (Alanko 1989).

For solution the Gauss-Seidel algorithm is written using two vectors (k+1 is the iteration cycle under process, N the number of equations and M the number of [linear] parameters)

$$t_i = -F_{hi} + \sum_{j=i+1}^{M} G_{ij} \cdot p_j^{(k)} \qquad i = 1, 2, \dots, N$$

$$z_i = \sum_{j=1}^{i-1} G_{ij} \cdot p_j^{(k+1)} \qquad i = 1, 2, \dots, N$$

in the form

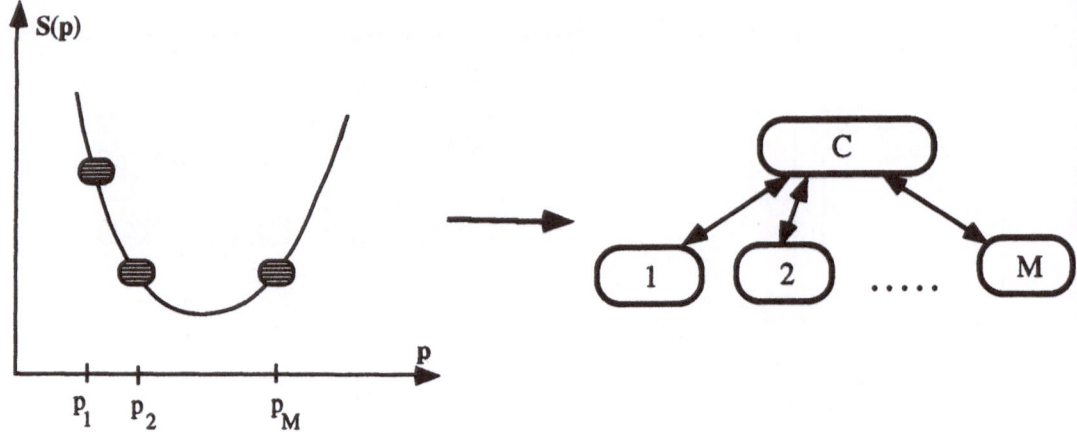

Fig. 9.5. Model optimization in parallel mode: the control processor C keeps track of the best objective function, which is computed in a different processor for each search for a value of p. (After Alanko 1989)

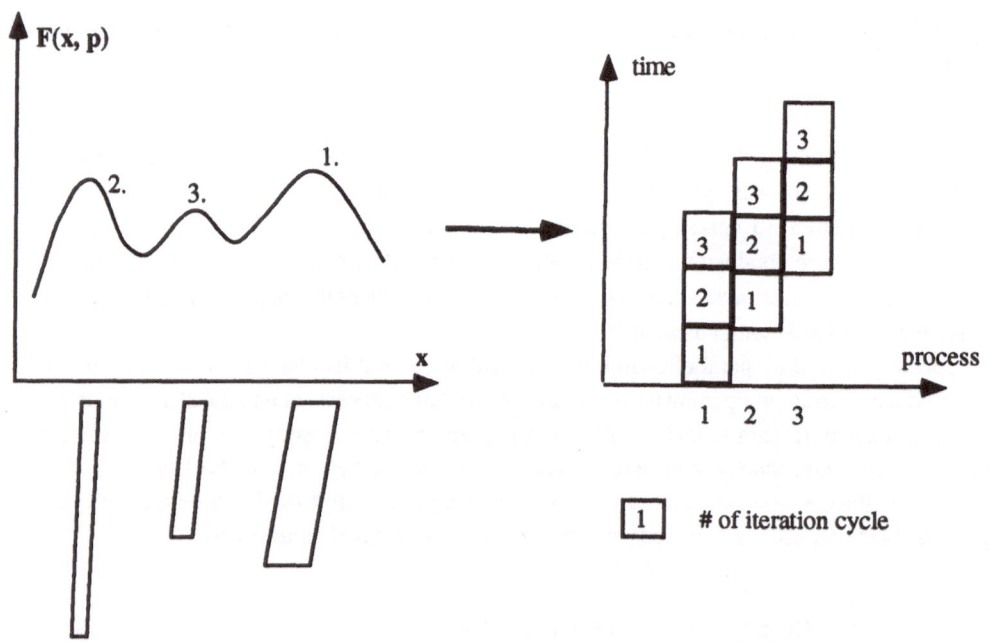

Fig. 9.6. Model optimization in parallel mode: several models are fitted to the data using the concept of partial anomalies (Hjelt 1973). One processor is dedicated to each model and the start of a new iteration of model j is delayed to have available the result of the most recent iteration of the model j - 1. (After Alanko 1989)

$$p_i^{(k+1)} = - (t_i^{(s)} + z_i^{(s+1)})/G_{ii}$$

For each parameter one needs to update only the vector z_j, since the value of t_i is obtained already during the preceding iteration cycle. During the first cycle, also the components of t can be obtained in parallel mode, if necessary.

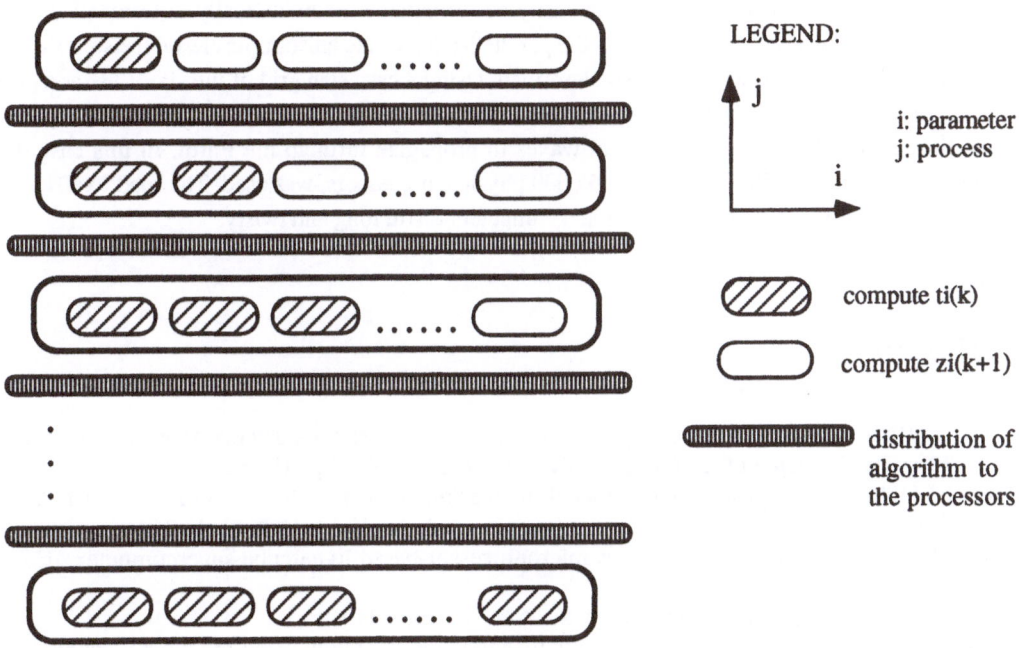

LEGEND:

j

i: parameter
j: process

i

compute ti(k)

compute zi(k+1)

distribution of
algorithm to
the processors

Fig. 9.7. Model optimization in parallel mode: the Gauss-Seidel iterative algorithm is distributed among the processors. The updated value of parameter i takes place in the process j. A complete iteration step is shown. (After Alanko 1989)

9.4 LARGE SYSTEMS OF EQUATIONS

Mora (1990) discusses modern trends of seismic inversion, making a plea for a unifying view, where both the high-wave number approach of seismic migration and the low wave number approach of tomography should be combined. At present, the greatest obstacle is the traditional way of sequential computing in one central processing unit. Recent developments in computer technology have made massively parallel machines feasible, in which tens of thousands of processors are operating simultaneously in a true parallel mode of computing.

Mora makes the case for parallelism of (geo)physical systems by describing the vibration of molecules in a room. Physics does not work by allowing one molecule to vibrate, then another and so on. Instead the molecules vibrate simultaneously all over the room making the physics of the air molecules in the room intrinsically parallel over the whole space domain. Time progresses sequentially and the system of molecules evolves according to the laws of nature. This feature makes it appealing to transfer the mathematics of a (geo)physical system into a computer system with many processors. Each particle, string, molecule or spatial location is handled in its own processor and only appropriate communication between them need to be secured by the programmer.

Mora (1990) points out that despite the intrinsic parallelism of a direct modelling problem, the inverse problem may not have similar parallel properties. It is, however, easy to find counter examples and geophysical inversion can easily be formulated mathematically in terms of one or several parallel forward problems. Briefly the parallelization of the anisotropic elastic (seismic) wave equation in three dimensions can be so described. Using a cartesian grid of the size 128^3 at a grid spacing of 110 km to embed the whole Earth, the calculation of the 3D seismic wave took about the same time as would have taken the real waves to propagate through the Earth. In this case the Connection Machine (Hillis 1987), with 65 000 physical processors, was used. The road to efficient and practical inversion on massively parallel computers is still long and twisty.

REFERENCES

Alanko I (1989) On the application of parallel processing in computer aided teaching of geophysics. MSc Thesis, Dept of Geophysics, University of Oulu, 79 + 7 pp (in Finnish)

Baker LJ (1989) Is 3-D wave-equation modelling feasible in the next ten years? In: Eisner E (ed): Supercomputers in seismic exploration. Pergamon Press, Exeter, UK, pp 1 - 10

Hayes JP, Mudge T, Stout QF (1986) A microprocessor-based hypercube supercomputer. IEEE Micro Oct 1986: 6 - 17

Hillis WD (1987) The connection machine. Sci Am 256: 108 - 115

Hjelt S-E (1973) Experiences with automatic magnetic interpretation using the thick plate model. Geophys Prosp 21: 243 - 265

Hjelt S-E (1989) Parallel processing and geophysical inversion (some abstract thoughts). Paper presented at the NOFTIG Meeting, Uppsala, Sweden, 11. - 13. 1.1989. (unpublished)

Hjelt S-E, Alanko I, Pirttijärvi M (1991) Parallel processing and geophysical inversion in micro-computerenvironment. In: Kananen H (ed): XV Natl Conf Geophysics, Oulu, 14.-15.5. 1991, pp 89 - 94

Hockney RW, Jesshope CR (1981) Parallel conputers. Adam Hilger Ltd, Bristol, 423 pp

Hwang K, Briggs FA (1986) Computer architecture and parallel processing. McGraw-Hill Int Editions, New York, 846 pp

Karp AH (1987) Programming for parallelism. Computer 20: 43 - 58

Kung SY, Lo SC, Jean SN, Hwang JN (1987) Wavefront array processors - concept to implementation. IEEE Computer July 1987: 18 - 32

Miranker WL (1971) A survey of parallelism in numerical analysis. SIAM Rev 13: 524 - 547

Moorehead WD (1989) Hypercube supercomputing. In: Eisner E (ed): Supercomputers in seismic exploration. Pergamon Press, Exeter, UK, pp 159 - 183

Mora P, Tarantola A (1989) Large scale elastic wavefield inversion. In: Eisner E (ed): Supercomputers in seismic exploration. Pergamon Press, Exeter, UK, pp 184 - 202

Mora P (1990) A unifying view of inversion. In: Desaubies Y, Tarantola A, Zinn-Justin J (eds): Oceanographic and geophysical tomography. NATO Adv Study Inst, Session L, 9.8. - 3.9. 1988. North-Holland, Amsterdam, pp 345 - 374

Myczkowski J, McCowan D, Mufti I (1991) Finite-difference seismic modeling in real time. Geophysics: The leading edge of exploration. 10 (6): 49 - 52

Ortega JM, Voigt RG (1985) Solution of parallel differential equations on vector and parallel computers. SIAM Rev 27: 149 - 213

White RE (1987) Multisplitting and parallel iterative methods. Comput Meth Appl Mech Eng 64: 567 - 577

SUBJECT INDEX

Algebraic reconstruction technique 110
Algorithm for parallel computation 197, 252
Analogue modelling 155
Analytic continuation 4, 222, 224, 225, 226
- inversion 8, 197
-- EM double-dipole systems 207, 210
-- example 9
-- static magnetic field data 200
Anomaly types 15
-, min-max 15
-, maximum 15
-, superposition of 15, 16, 18, 19
Apparent resistivity 41, 91, 97, 98
A posteriori covariance matrix 191
-- probability 182, 187, 188
A priori covariance matrix 187
-- probability 182, 187, 191
Attractor basin 136, 137

Backus and Gilbert
- approach 101
- averaging function 103
Basin of attraction 138, 139
Bayesian estimation 184, 186
Bedrock interface model 74, 77, 82
Bit 179
Block model 13
--, effect on computing time 27
--, minimum distance in Gauss-Seidel
    iteration 62
Borehole magnetometry 201, 206
- radar method 105
-, role in tomography 105

Centre of probability density 177, 178
Characteristic points (= special points) 4, 6
--, accuracy of inversion 46
-- combined measures 35, 36
--, example of determination 37, 38, 42
-- horizontal measures 35, 36
-- vertical measures 35, 36
Combination of field components 197
Complex interpretation 3

- representation of static magnetic fields 200
Condition number of a matrix 62
Conductivity
-, excess 55, 216, 219, 221
-, piecewise continuous 56
Conductor
-, cylindrical 217, 218
-, half-plane 43, 207, 208, 209
-, inclusion 220, 221, 225
Conjugate directions 150
--, methods 144, 145, 150
Constrained optimization 127, 128, 185
Constraints of parameters (error analysis) 234
- in statistical distribution 179, 185
Continuation
-, downward 4, 222
-, example 224, 225
- of static potential fields 222
-, upward 4, 222
- of wave fields (migration) 226
Correlation
- between parameters 20, 123, 125, 234
--, magnetic example 238, 239
- coefficient as a measure of fit 17
-, effect on parameter errors 234
-, effect on constraining parameters 237, 239
- matrix 178, 183
Coupled eigenvalue equations 73
Covariance matrix 178, 241
-, a posteriori 188, 191
-, a priori 187, 241
-- operator 178
Curse of dimensionality 140
Cramer-Rao bound 233, 234, 237
Cylinder model
-, geoelectric 217
-, horizontal, gravity field 46

Data point density, example 21
- eigenvector 73
- space 64
- vector 58
--, generalized 65, 67
Databank inversion 44, 45
Davidon method of optimization 161, 162

DC sounding (Schlumberger) 97
Density contrast 165
Determinant of a matrix 62
Deviation
-, mean 178
-, standard 178
Diagonally dominant matrix 61
Digit, measure of information 179
Dimensionless parameters in nomograms
   35, 43
Dimensionality
-, curse of 140
- of models 3
Dipole EM systems
-, airborne 207
-- dipole-dipole geometry 207, 208
Direct search methods 141
Double-dipole EM systems 207, 208, 209

Eigenvalue 64, 65
- equation 64, 70, 73
- matrix 65
- of a matrix 64
- of a rectangular matrix 72
- of a square matrix 64, 65
-, shifted 72
-, small, effect of 69, 71
-, zero 66, 67
Eigenvector 64, 65, 73
-, data 73
-- example 94, 95
- normalization 69
- of a matrix 64
- of a rectangular matrix 72
- of a square matrix 64, 65
- orthogonality 72
-, parameter 73
-- example 94
Entropy 180, 181
-, method of maximum 181
- of a continuous signal 181
Equivalence 22, 23, 24, 25
Error analysis 231
-, classification of 231
- covariances 237, 242
-, examples for magnetic plate model 202,
   203, 234, 235, 236, 237, 238, 239
-, linearized 232
- of parameters 233
--, minimal 233, 234

-, random, effect of 202, 203
- sensitivity 207
Estimation, Bayesian 184, 186
Exact fit 60, 118
Excess conductivity 216, 219, 221
Excitors of fields 216
Extreme models 13, 15
Extremum
-, global 121, 131, 140
-, local 121, 140

Fibonacci numbers 134
- search 133
Filtered backprojection 109
Filtering
-, high-pass, in upward continuation 222
-, low-pass, in downward continuation 223
- to change anomaly type 4, 5
Finite functions method 219
Fit
-, exact 60, 118
-, least-squares 17, 60, 118
-, measures of 17
-- correlation coefficient 17
-- M-fitting 18
-- weighted sum of squares 17, 189
-, minimax 17, 119
-, parabolic 134, 157
Functional approximation of model bodies
   219

Gaussian function 178
- noise 183, 184
- statistics 233
Gauss-Seidel method 61
-- use in parallel computation 251, 252
Generalized data vector 65
- inversion 64, 67
-- using Singular Value Decomposition
   (SVD) 64, 110
- parameter vector 65
Genetic algorithms 155
Geoelectric model 216
Golden cut 132, 133
Gradient methods 136
- method of optimization 149
--, numerical sensitivity 137, 149, 151
-, regional, in magnetometric inversion 157
Green's function for half-plane 207

Half-plane Green's function 207
-, infinitely well conducting 43, 207
--, analytic inversion 207
--, nomogram 44
Half-range, probability density function 178
Heuristic approach (in random search) 155
Hidden layer 23, 24
Horizontal-loop method 42
Hyperparabolic fit 45, 142, 156

Ill-conditioned matrices 153
Ill-posed problem 109, 222
Image dipole 207, 208
Imaginary component 42, 225
In-phase component 42, 220
Information
-, average 180
- content 179
- density matrix 57, 70, 92, 96
--, graphical exposition 75
- matrix 233
-, measure of 179
-, total 180
-, units of 179
Interactive inversion 7, 131
Interpretation (see inversion)
Inverse matrix 65
Inversion 3
-, analytic 197
- examples
-- gravity 8, 9, 46, 165
-- magnetometry 156
-- magnetotellurics (MT) 167
-- seismic refraction 166
-, factors affecting 18
-, generalized linear 64, 67
-, non-uniqueness of 20, 29, 30
- of a matrix 62
- operator 8, 54, 198, 201
-- for magnetometry, examples 198, 201
-, properties of methods 4
-, stability of 62
Iteration step in linearized inversion 56
Iterative inversion 7
-- examples 56, 129
-- of linear problems 61
--, starting values (initial guess) 61

Joint inversion 3, 4, 97

Lagrangian multipliers, use of 102, 103, 127, 128, 185, 186, 190
Lanczos inverse see Generalized linear inversion
Layer models
-- in DC sounding 97
-- in EM sounding 90
-- in joint inversion 97
-- in magnetotellurics 167
-- of density 78
-- with inclined boundaries 166
Least squares (LSQ) method 8, 184
-- inverse operator, example 8
--, recursive weighted 242
Levenberg-Morrison-Marquardt method (see Marquardt method) 152
Linear
- equation systems 57
--, determined 57, 58, 59
--, overdetermined, error analysis 57, 58, 59
--, underdetermined 57, 58, 59, 96
- parameters 51
--, genuine 51
--, intermediate 54
--, linearized 51
- problem 51
- programming 129
Linearization
- of objective functions 149
- of parameters 51, 54
LSQ (see least squares method)

Magnetometer array 197
Magnetometry
- analytic inversion, example 125, 200, 205
- comparison of optimization methods 156, 164
- maps of objective functions 121
- minimum parameter error 234, 235, 236, 237
- multiplate models 160
Magnetotellurics 130, 153, 154, 167, 197
Marquardt method 151, 156
-- examples of use 97, 157, 158, 161, 164, 166, 167
-- normalization 153
Matrix
-, condition number 62, 63
- determinant 62
-, diagonally dominant 61, 62
- eigenvalue 64, 65
- eigenvector 64, 65, 72

-, inversion of 62
- inverse 65
Maximum entropy method (MEM) 181, 184, 186
Maximum likelihood (ML), principle of 181, 184, 186
Mean deviation of probability density function 178
Mean of probability density function 178
Measure of characteristic points 35, 36
- of fit 17, 19
--, minimization of 118
Median of probability density function 178
Merging of anomaly peaks 83
Mid-range of probability density function 178
Migration of fields 226
-, electromagnetic 226
-, seismic 226
Minimax fit 17, 119
Minimum
-, global 121, 129, 131, 155
-, local 121, 129, 131, 155
Model
-, choice of 15
-, extreme 13, 15
- field 102
-, flexibility of 13
- parameter 5, 60
-, properties of 16
Monte-Carlo search 153
Multidimensional search methods 138, 141
- direct search methods 141
- sequential search 142
Multiparameter problems 131
Multipole expansion 215
Multivariate search in optimization 141
- conjugate directions 144, 150
- hyperparabolic search 142
- pattern search 144
- random search methods 153
- simplex method 147
- steepest descent methods 149, 150

Nep 179
Newton's method 136
-, basin of attraction 138, 139
-, numerical sensitivity of 137
Nomogram 6
-, computerized 44
-, construction of 35
- dimensionless parameters 35, 43
-, examples of 38, 41, 42

-, use of 39, 40, 41, 44
Non-linear equations 117
- objective functions 121
- optimization 117
--, constrained 127
-- multidimensional methods 141
--, one-dimensional 130
--, stopping criteria 129, 152
-- with variable strategy 155, 165
Non-linearity, degree of 121
Non-uniqueness 20, 29, 30
Norm 117
-, L_1 118
-, L_2 118
-, LSQ 118
-, L_n 118
-, Minimax 119
- of a parameter function 102
Normal distribution of errors 187
Normalization of eigenvectors 69
Number of function determinations 131
Numerical sensitivity of
- gradient methods 137, 149, 151
- Newton's method 137

Objective function 117
--, choice of 157
--, convexity of 122
--, maps of 121, 122, 124, 125, 126
--, multimodality of 121, 122
--, properties of 121, 156
--, unimodality of 121, 122, 140
Occam's razor 96
One-dimensional search (for non-linear optimum) 130, 131
- Fibonacci search 133, 134
- golden cut 132, 133
- gradient method 136, 137
- Newton's method 137
- parabolic fit 134, 135
- secant method 136
Operator, inversion 8, 54, 198, 201
Outlier data 119
Out-of-phase component 42, 220

Parabolic fit 134
- example of use 157
Parallel computers
- architecture 247

--, massively 28, 254
-- topology 247
Parallelism
- effect on anomaly calculation 28
- in computation, seismology 248, 254
- in data 248
- in optimization 155, 250
- of algorithms 247
Parameter
-, continuous function 101
-, intermediate 54
-, linear 51, 60
-, minimum length of 69
-, non-linear 53
-, piecewise constant 52, 56
- resolution 103, 131
- space 64
--, vasteness of 138, 141
- trade-off 103
- variance 58, 69
- vector, examples
-- in DC sounding 197
-- in electromagnetics 43, 90, 97
-- in gravity 46, 76
-- in magnetometry 54, 83, 157 et seq.
-- in magnetotellurics 167
-- in refraction seismics 166
-- in VLF resistivity 41
Parametrization of model body contour 219
Partial anomalies (profiles) 156
--, method of 156
-- in parallel computation 156, 251, 252
--, interaction of 83
- matrix 58
Pattern recognition 30
- search 144, 146
Penalty function 127
--, exact 127
--, sequential 127
Plate model
-, gravity, nomogram 39, 40
-, magnetic
-- analytic inversion 125, 200, 205, 206
-- error analysis 234
-- maps of objective function 121
-- minimum parameter errors 234, 235, 236, 237
-- parameter correlation 236, 237
-- SVD analysis 82
--, thin plate approximation 53, 62, 202
Polygonal model, 2D in gravimetry 165
Powell's method of optimization 151, 156,
    158, 159, 161, 162

Probabilistic formulation of inversion 187
Probability
-, a posteriori 182, 191
-, a priori 182, 190
-, conditional 182, 187, 191
- density function 177, 190
--, centre of 177, 178
--, dispersion of 177, 178
--, mean of 178
--, mean deviation of 178
--, median of 178
--, mid-range of 178
--, standard deviation of 178
Profiling 3

Quadratic programming 128
Qualitative inversion 3
--, aims 5
--, pitfalls 5
Quantitative inversion 4, 5

Radon transform 109
--, inverse 109
Random search methods (see statistical
    methods of optimization)
Rank of a matrix 58
Real component 42, 225
Regional background field 165
Regula falsi 136
Regularization 223
- error analysis 240
- example in continuation of fields 216, 217
Resistivity, apparent 41, 42, 91, 98, 99
Resolution
- matrix 57, 70
-- examples 71, 75, 80, 92, 95, 103
- of model parameters 102
Ridge regression 100
Robust methods (of non-linear optimization)
    119
Rosenbrock's method 144, 145
Rule of thumb technique 4, 6, 35

Scale models 6
Secant method 136
Seismic refraction, non-linear optimization
    166
- tomography 105, 110
- velocity 166

Sequential search 142
Simplex
    method 147, 156
- operations 147, 148
-- contraction 148
-- expansion 148
-- reflection 147, 148
- properties 147
-- centroid 147, 148
Simulated annealing 155
Singular points, method of 216
SVD (Singular value decomposition) 64
- of a rectangular matrix 72
- of a square matrix 66
- ridge regression variant 100
-, use of
-- in electromagnetism 90
-- in gravity 73, 78
-- in magnetometry 81, 82
-- in oceanography 96
-- in seismology 94
Slingram 42, 43, 44, 45
Slowness 106
Smoothness of parameter vectors 188
Sounding 3, 90, 97, 167
Source distributions 215
-, loop 56
-, seismic 105, 107, 108
Speedup of convergence in linear problems
    61
Spread of parameter function 103
Stabilization functional 216
Standard deviation of probability density
    function 178
Starting values, Gauss-Seidel iteration 62
Statistical methods of optimization
- genetic algorithms 155
- Monte-Carlo inversion 153
- simulated annealing 155
- tabu search 155
- tunneling algorithm 155
Steepest descent methods 149, 151
-, numerical sensitivity of 149, 151
Strip, infinitely well conducting 223
Stopping criteria of iteration 129, 152
Sum of squares 156

Tabu search 155
TEMS (transient EM sounding) 97
Tightening contours, method of 216, 218
Tomography 104

- regional seismic study 110
-, resolution in 106
Topology of processors 247
Trade-off of parameter variance and reso-
    lution 103, 104, 106, 241
Transformation of data 5
Travel path of seismic waves 105, 106
- time of seismic waves 105, 106
Tunneling algorithm 155

Variance of inversion 233
-, parameter, minimum of 58
Vastness of parameter space 138, 141
Vector of data 58, 64
- of model parameters 58, 64
VLF resistivity method 41, 42

Weighted
- LSQ 17, 189
--, recursive 242
- objective function 119
Werner deconvolution 53

Lecture Notes in Earth Sciences

Vol. 1: Sedimentary and Evolutionary Cycles. Edited by U. Bayer and A. Seilacher. VI, 465 pages. 1985. (out of print).

Vol. 2: U. Bayer, Pattern Recognition Problems in Geology and Paleontology. VII, 229 pages. 1985.

Vol. 3: Th. Aigner, Storm Depositional Systems. VIII, 174 pages. 1985.

Vol. 4: Aspects of Fluvial Sedimentation in the Lower Triassic Buntsandstein of Europe. Edited by D. Mader. VIII, 626 pages. 1985.

Vol. 5: Paleogeothermics. Edited by G. Buntebarth and L. Stegena. II, 234 pages. 1986.

Vol. 6: W. Ricken, Diagenetic Bedding. X, 210 pages. 1986.

Vol. 7: Mathematical and Numerical Techniques in Physical Geodesy. Edited by H. Sünkel. IX, 548 pages. 1986.

Vol. 8: Global Bio-Events. Edited by O. H. Walliser. IX, 442 pages. 1986.

Vol. 9: G. Gerdes, W. E. Krumbein, Biolaminated Deposits. IX, 183 pages. 1987.

Vol. 10: T.M. Peryt (Ed.), The Zechstein Facies in Europe. V, 272 pages. 1987.

Vol. 11: L. Landner (Ed.), Contamination of the Environment. Proceedings, 1986. VII, 190 pages.1987.

Vol. 12: S. Turner (Ed.), Applied Geodesy. VIII, 393 pages. 1987.

Vol. 13: T. M. Peryt (Ed.), Evaporite Basins. V, 188 pages. 1987.

Vol. 14: N. Cristescu, H. I. Ene (Eds.), Rock and Soil Rheology. VIII, 289 pages. 1988.

Vol. 15: V. H. Jacobshagen (Ed.), The Atlas System of Morocco. VI, 499 pages. 1988.

Vol. 16: H. Wanner, U. Siegenthaler (Eds.), Long and Short Term Variability of Climate. VII, 175 pages. 1988.

Vol. 17: H. Bahlburg, Ch. Breitkreuz, P. Giese (Eds.), The Southern Central Andes. VIII, 261 pages. 1988.

Vol. 18: N.M.S. Rock, Numerical Geology. XI, 427 pages. 1988.

Vol. 19: E. Groten, R. Strauß (Eds.), GPS-Techniques Applied to Geodesy and Surveying. XVII, 532 pages. 1988.

Vol. 20: P. Baccini (Ed.), The Landfill. IX, 439 pages. 1989.

Vol. 21: U. Förstner, Contaminated Sediments. V, 157 pages. 1989.

Vol. 22: I. I. Mueller, S. Zerbini (Eds.), The Interdisciplinary Role of Space Geodesy. XV, 300 pages. 1989.

Vol. 23: K. B. Föllmi, Evolution of the Mid-Cretaceous Triad. VII, 153 pages. 1989.

Vol. 24: B. Knipping, Basalt Intrusions in Evaporites. VI, 132 pages. 1989.

Vol. 25: F. Sansò, R. Rummel (Eds.), Theory of Satellite Geodesy and Gravity Field Theory. XII, 491 pages. 1989.

Vol. 26: R. D. Stoll, Sediment Acoustics. V, 155 pages. 1989.

Vol. 27: G.-P. Merkler, H. Militzer, H. Hötzl, H. Armbruster, J. Brauns (Eds.), Detection of Subsurface Flow Phenomena. IX, 514 pages. 1989.

Vol. 28: V. Mosbrugger, The Tree Habit in Land Plants. V, 161 pages. 1990.

Vol. 29: F. K. Brunner, C. Rizos (Eds.), Developments in Four-Dimensional Geodesy. X, 264 pages. 1990.

Vol. 30: E. G. Kauffman, O.H. Walliser (Eds.), Extinction Events in Earth History. VI, 432 pages. 1990.

Vol. 31: K.-R. Koch, Bayesian Inference with Geodetic Applications. IX, 198 pages. 1990.

Vol. 32: B. Lehmann, Metallogeny of Tin. VIII, 211 pages. 1990.

Vol. 33: B. Allard, H. Borén, A. Grimvall (Eds.), Humic Substances in the Aquatic and Terrestrial Environment. VIII, 514 pages. 1991.

Vol. 34: R. Stein, Accumulation of Organic Carbon in Marine Sediments. XIII, 217 pages. 1991.

Vol. 35: L. Håkanson, Ecometric and Dynamic Modelling. VI, 158 pages. 1991.

Vol. 36: D. Shangguan, Cellular Growth of Crystals. XV, 209 pages. 1991.

Vol. 37: A. Armanini, G. Di Silvio (Eds.), Fluvial Hydraulics of Mountain Regions. X, 468 pages. 1991.

Vol. 38: W. Smykatz-Kloss, S. St. J. Warne, Thermal Analysis in the Geosciences. XII, 379 pages. 1991.

Vol. 39: S.-E. Hjelt, Pragmatic Inversion of Geophysical Data. IX, 262 pages. 1992.